Mayas & Aliens

MOHAMED CHERIF

DEDICATION

My dedications to the truth seekers around the world whatever their origins, colors, ethnicities, religions, etc.

CONTENTS

CHAPTER I: ALIENS ARTIFACTS

Before the last Mayan calendar finished on December 21st, 2012 and new one started for another 5126 year, a special documentary was prepared for this event showing the world the new discovered artifacts which were conserved for a while by the Mexican government.

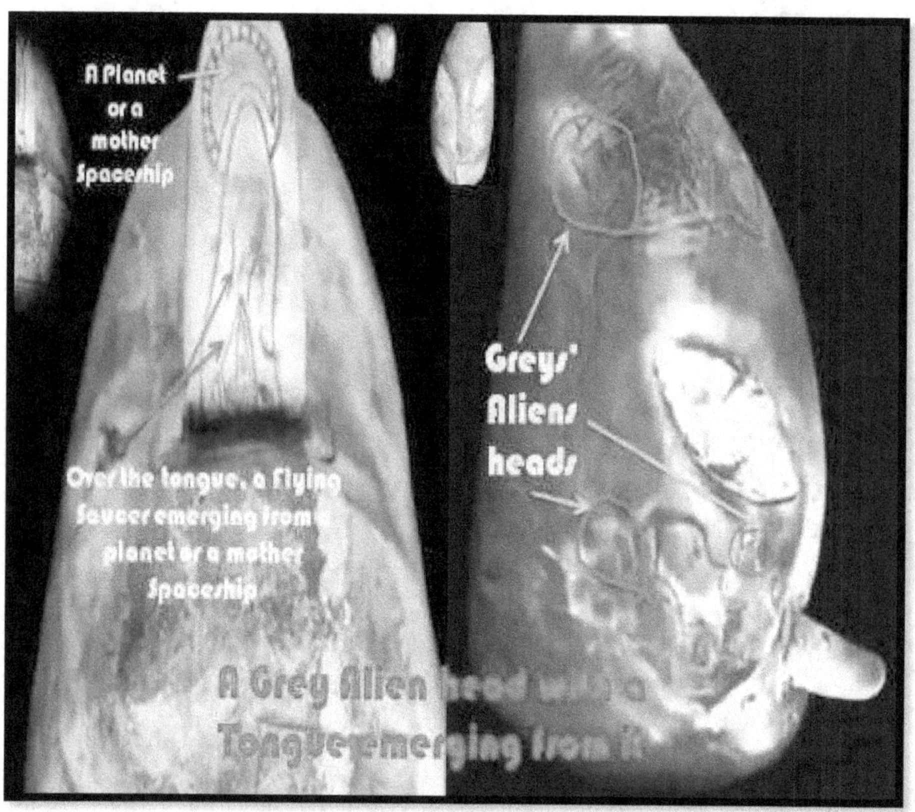

This documentary was produced by Raul Julia-Levy in collaboration with the Mexican ministry of tourism presented by Luis Augusto Garcia Rosado.Mr Rosado declared in a statement : "new evidence has emerged of

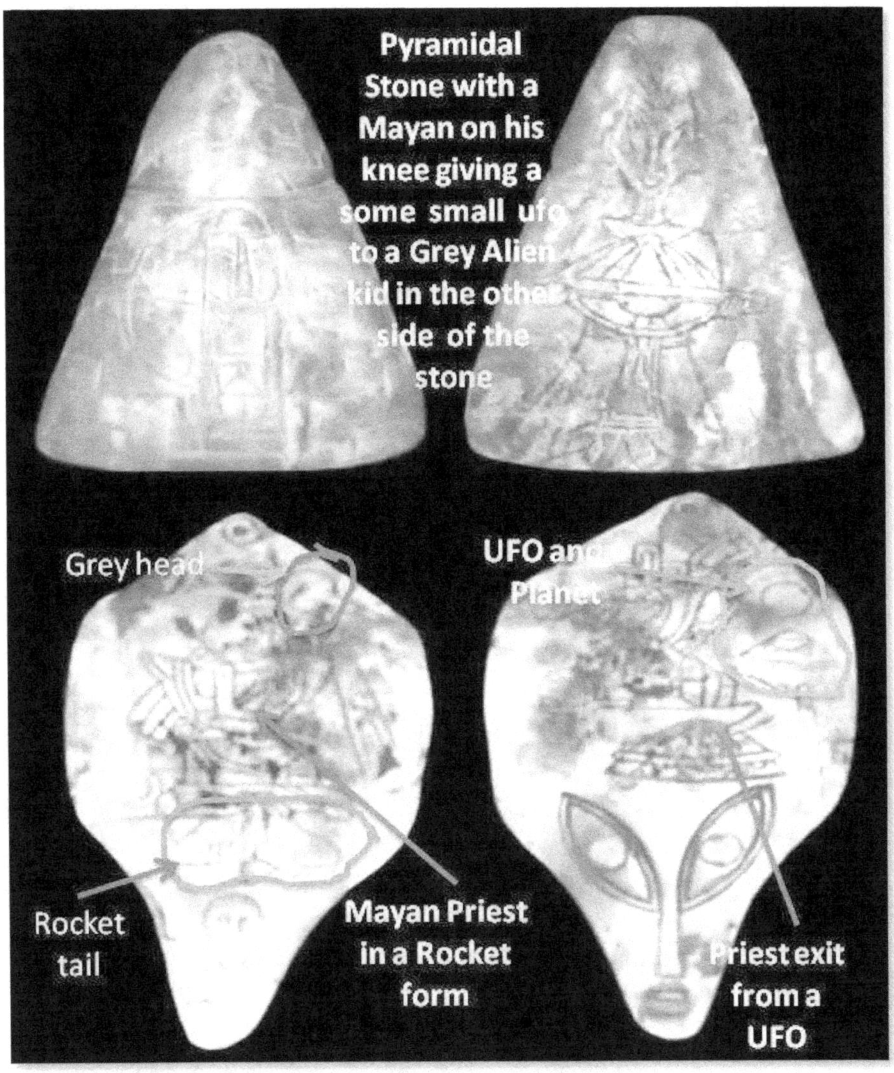

contact between the Mayans and extraterrestrials, supported by translations of certain codices (he didn't mentioned which one), which the government has kept secure in underground vaults for some time." lately he replied in a communication:"landing pads in the jungle that are 3,000 years old». He said too that his country was simply offering the filmmakers' access to previously unexplored sections of a Mayan site at Calakmul.

The same event has interesting the Guatemalan ministry of tourism to make documentary about this subject as the discoveries were too in Guatemala .Guatemala is the site of a large number of pre-Columbian Mayan settlements in the Mirador Basin including the extensive and highly organized city of El Mirador, Mirador is not the name of a pyramid; it's the name of the entire settlement, which includes several pyramids, the largest one is LaDanta.

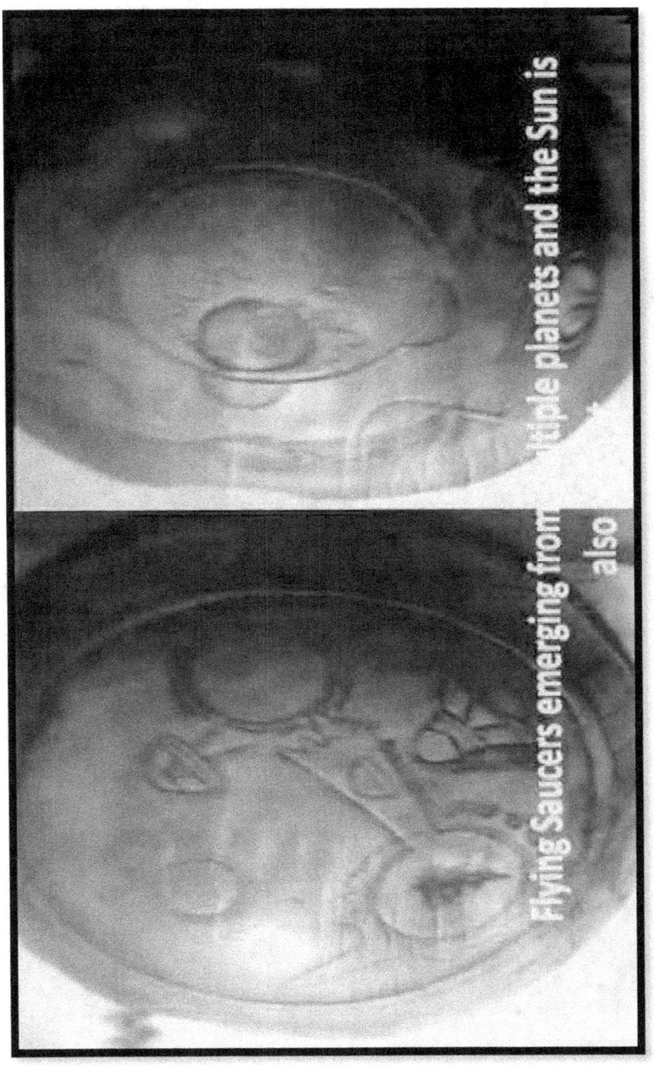

It's clear that all the concentration of officials in the two countries was the benefits of tourism behind the extraterrestrials artifacts something which can make doubts about the credibility of discovering and their real age. Artifacts were shown to international archeologists whose maybe reflects the contrary.

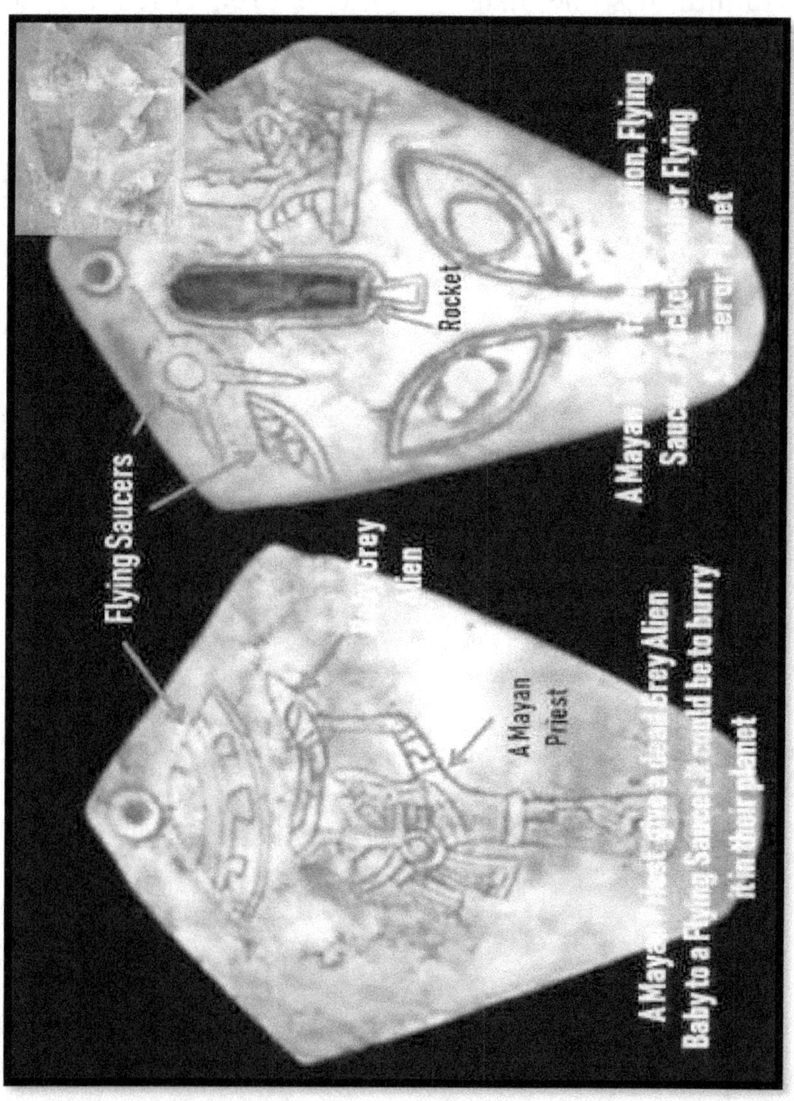

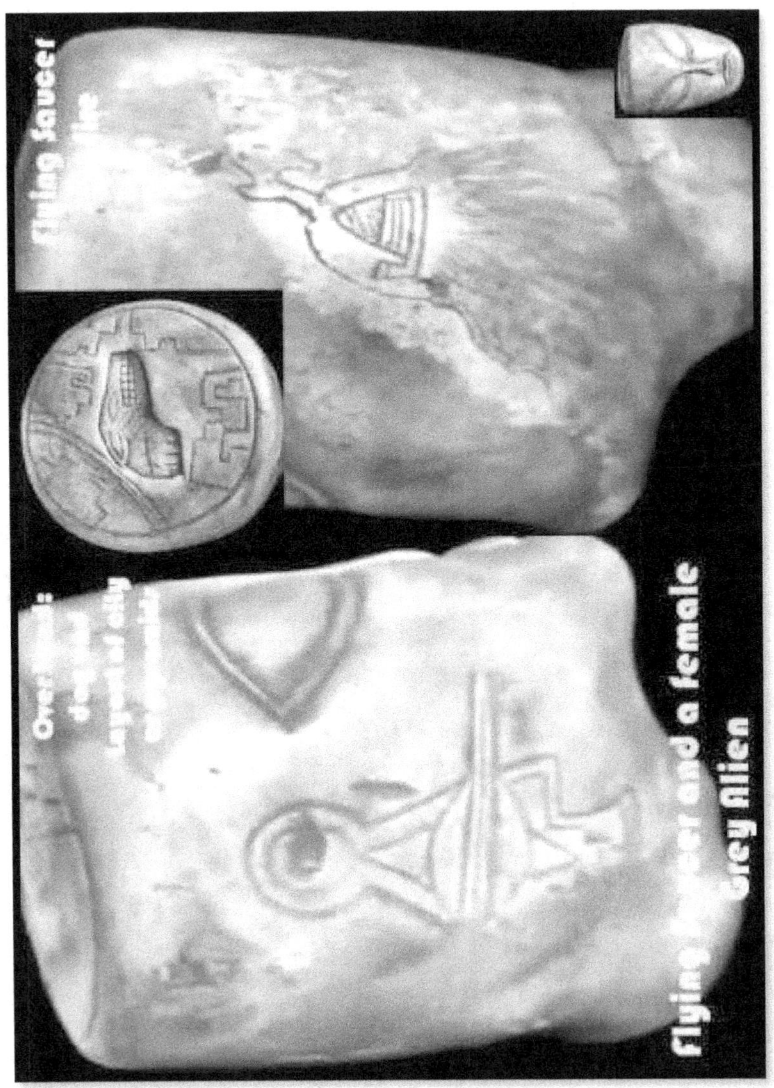

The producer Raul-Julia claims there is proof that the Mayans had intended to lead the planet for thousands of years, but were forced to escape after an invasion by "men of dark intentions," leaving behind evidence of an advanced race.

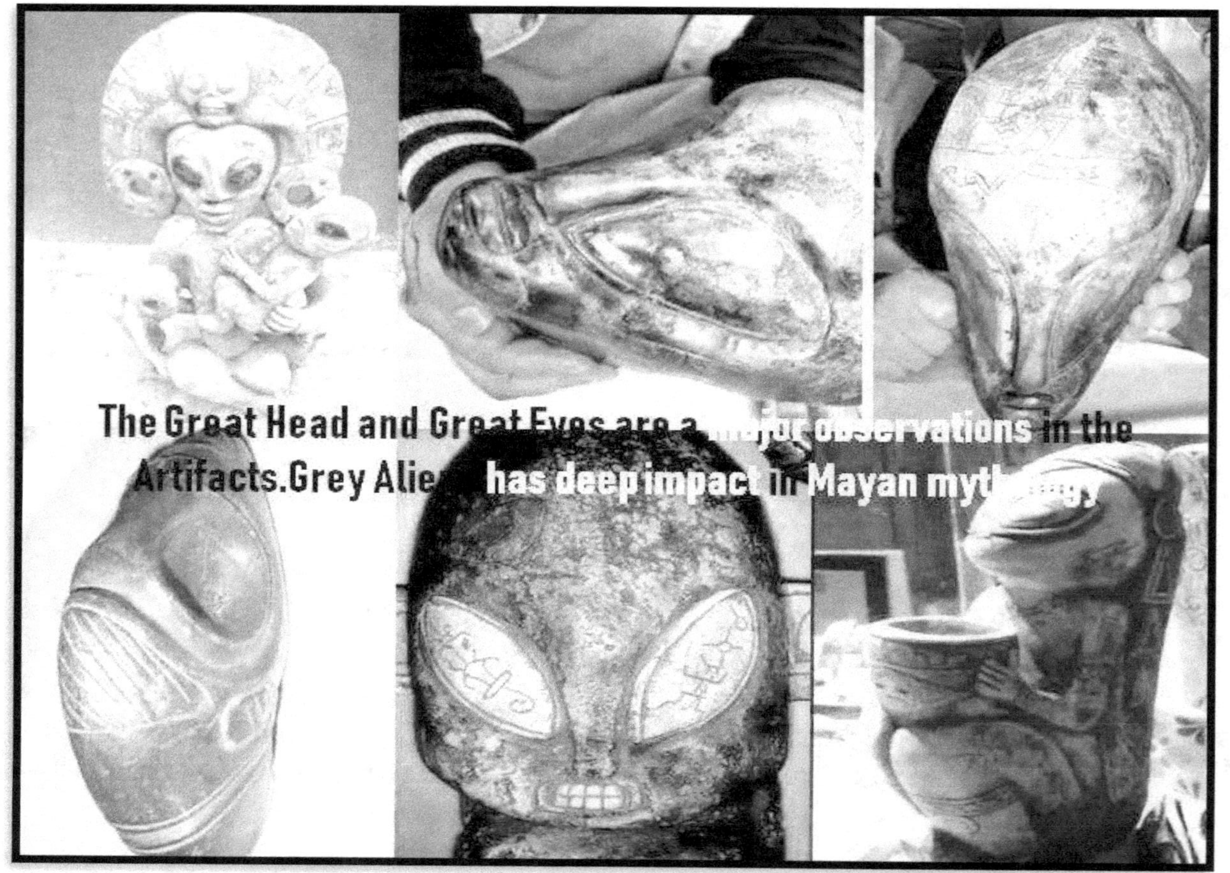
The Great Head and Great Eyes are a major observations in the Artifacts.Grey Alien has deep impact in Mayan mythology

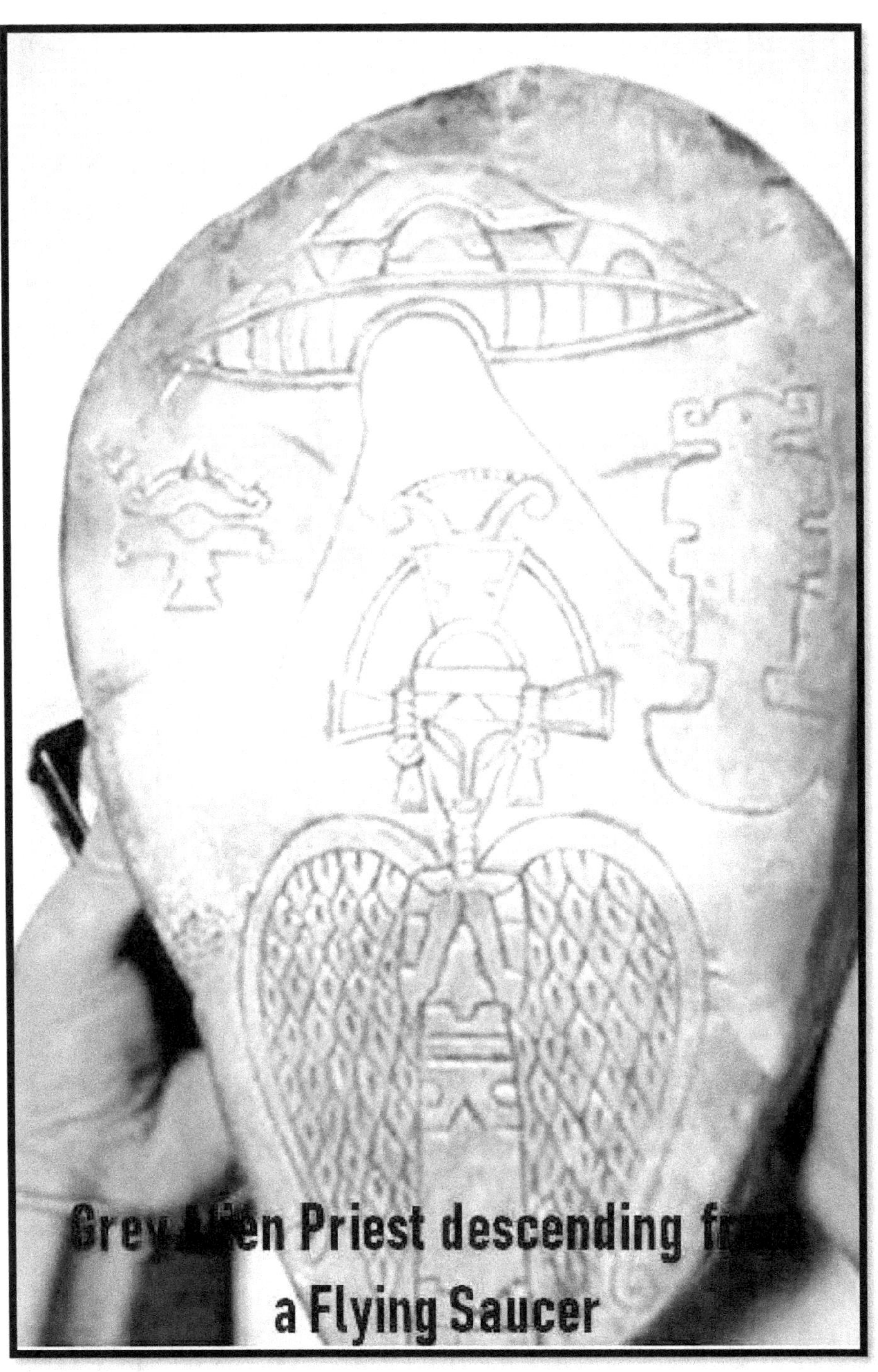

Grey Alien Priest descending from
a Flying Saucer

Exchange of gifts between
Mayan Priests and Grey Aliens
in both sides of the mask

Mayan Priests

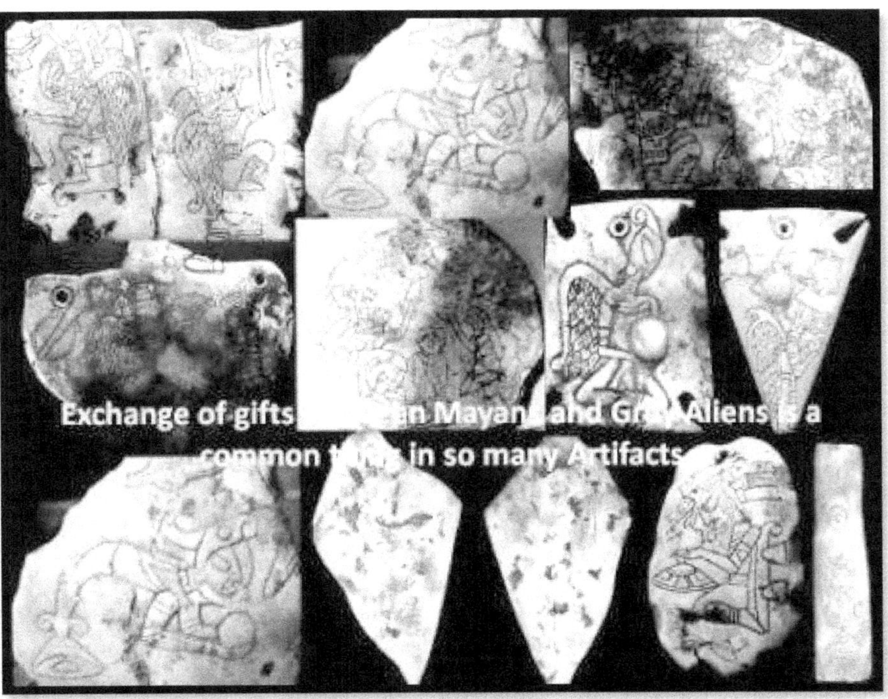
Exchange of gifts between Mayans and Grey Aliens is a
common thing in so many Artifacts

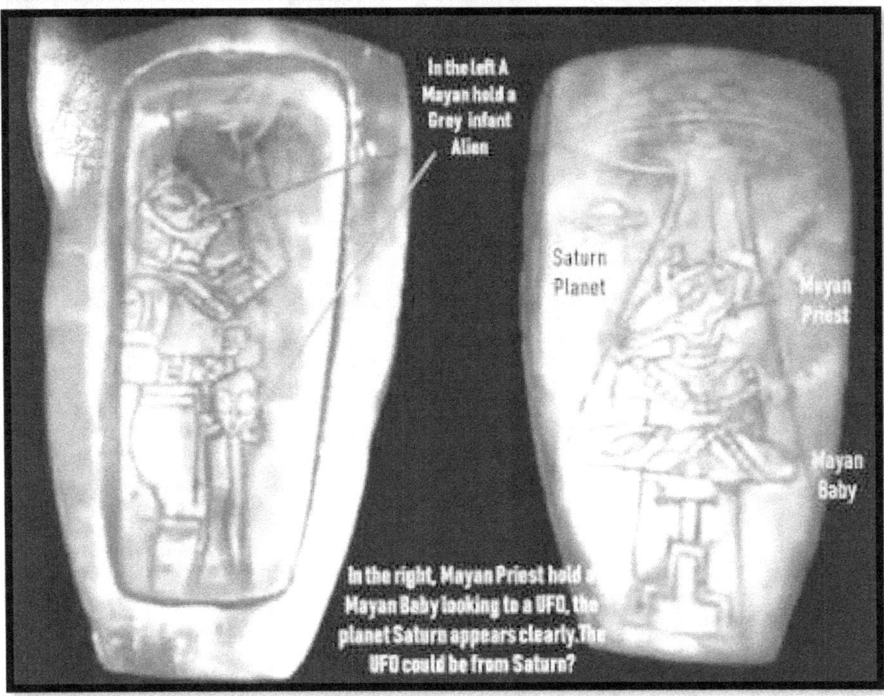
In the Left A
Mayan hold a
Grey Infant
Alien

Saturn
Planet

Mayan
Priest

Mayan
Baby

In the right, Mayan Priest hold a
Mayan Baby looking to a UFO, the
planet Saturn appears clearly. The
UFO could be from Saturn?

Flying Saucers taking off from a planet and travelling with an
Alien wearing an astronaut clothes inside the cockpit

A Mayan Priest hold a dead Grey Alien.The serpent is a symbol of transfer to the death surrounding a Grey head
A Grey Alien emerging from a Flying Saucer

CHAPTER II: CODICES

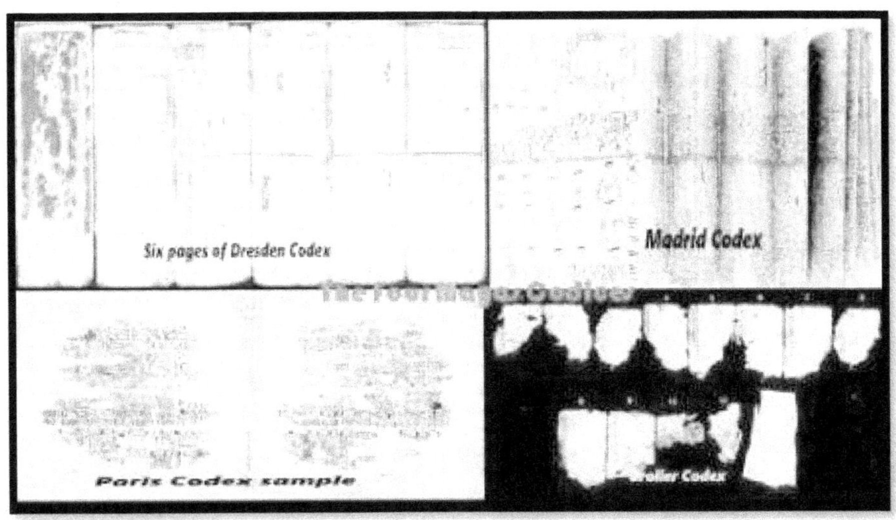

The four codices plural of codex are belonging in their names for the city which where they exists now: Dresden Codex in Germany, Paris Codex in France, Madrid Codex in Spain and Grolier Codex in New York Grolier Club.

The Dresden Codex was written mostly between the twelfth and fourteenth centuries in paper's both sides, composed by 39 pieces forming 74 pages with length as 3.56 meters (11.7 feet).It took his way to Europe, Germany in 1739.More than sixty per cent of this Codex presents rich glyphs of astronomical tables focusing on eclipses, equinoxes, solstices and description of planet cycles like Mars, Venus and the Moon. These tables predicted accurately astronomical events for 33 years in the end of classic period.

The Madrid Codex was acquired from Spanish Museum in 1872 from a Collector. It contains 112 pages; 56 sheets written both sides. It's the longest preserved Mayas Codices. A number of Priests wrote its content. It consists almanacs and horoscopes which were used to help them in their ceremonial rituals. It contains too astronomical tables.

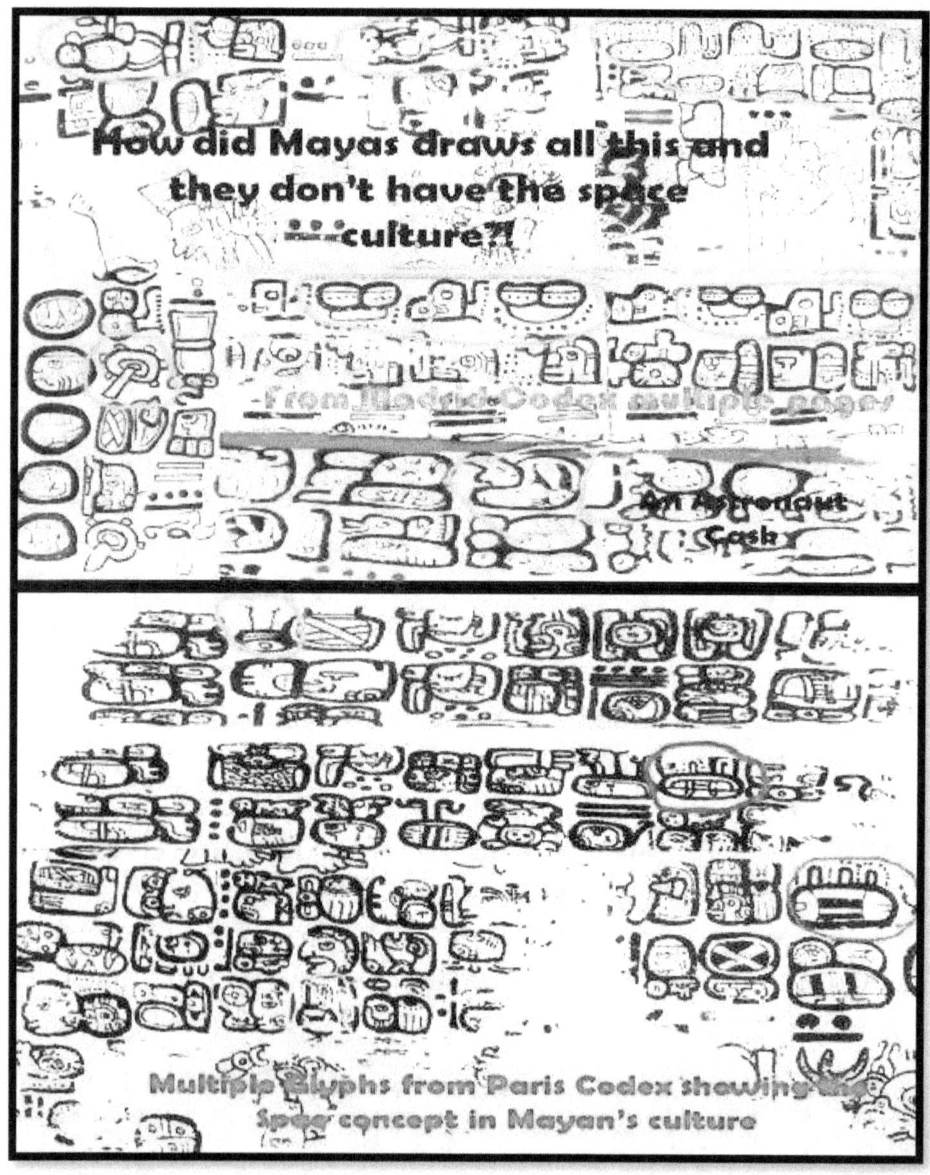

The Paris Codex is formed by 22 pages painted both sides with a length of 140 centimeter (55 inches).It was discovered in Paris in 1859 after being classified in 1832 without being noticed by Juan Pio Perez. Mostly consisting about the post classic Mayans period from the eighth to the tenth centuries. The western of Yucatan peninsula was the holder of this Codex. It contains rituals with relation of calendars, astronomical signals and animals representing a zodiac.

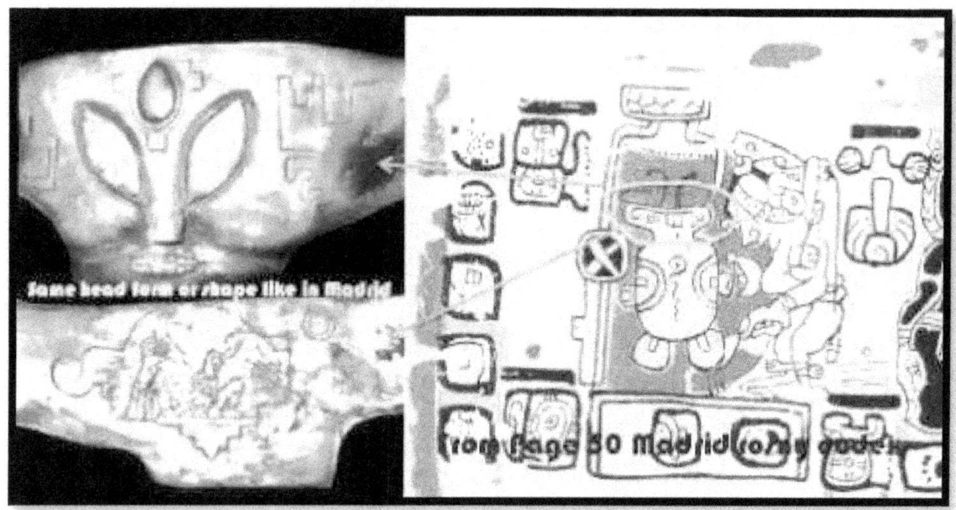

The Grolier Codex, this name came from the Ownership of the Grolier Club of this Codex, was recognized just in the twentieth century precisely in 1970.It was found in a cave by a Mexican collector and donated it to the Mexican Museum of Anthropology. It contains 11 pages most of

them can't give a clear description for any kind of activity of the Mayans. Most of pages show a priest or king who is cutting a neck of a captive which represents a sacrifice for the god which can be a planet of Venus or Mars or make another ritual in a Mayan citizen. The page ten is approximately destroyed by ninety per cent. It was just in 2018 its authenticity was confirmed by the Mexican Official Archeologists.

CHAPTER III: MAYAN CALENDARS

To study the sky, Mayas built observatories in many large cities; the most standing up one is the Chichen Itza observatory. The priests studied the movements of the sun, moon, planets and stars. They could predicts so many sky events like Eclipses, moons phases, planets, etc .Every moving object which appeared and disappeared in recurrent style with periodic manner was considered as a god and this god is deeply in relation with earth events and the societies living on it.

The priests used their deep knowledge of Astronomy to develop accurate calendars, they had two types of calendars, one was a sacred calendar and the other was used for planning regular events. The sacred calendar, called the Tzolkin had 260 days. It used 13 cycles of twenty days names and each day had a particular god with its unique symbol associated with it. They did not divide this sacred calendar into months. The Mayas used this calendar to determine religious events and naming children. A 365 days calendar was also used, it has 18 months each month has twenty days, it give 360 days the five days are considered the unlucky ones.

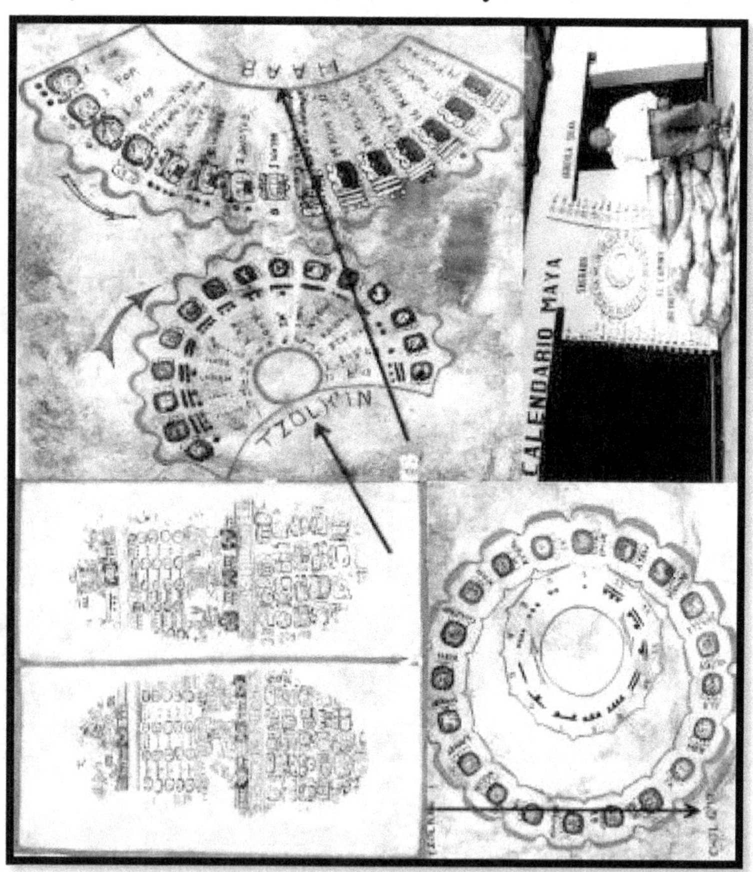

CHAPTER IV: MAYA'S ASTRONOMICAL CHARTS

From the jungles of Guatemala, the oldest Mayan's astronomical charts until now were discovered by group of specialists from Texas University. Dating from the ninth century, the charts were hidden in a room walls under rainforest vegetation overgrown an ancient building in Xultun in the north-east of Peten.

One wall was covered with hundreds of small red and black symbols that tracked the phases of the moon, with others representing the Mayan ceremonial calendar and cycles of the sun, Mars and Venus.

"It's the oldest Mayan astronomical table. It's the only one

discovered from the Mayan's classical period. It's also the first time looking inside an astronomer's house and see the writing on the wall," declared one of the excavation team. More enigmatic charts on the north wall appear to

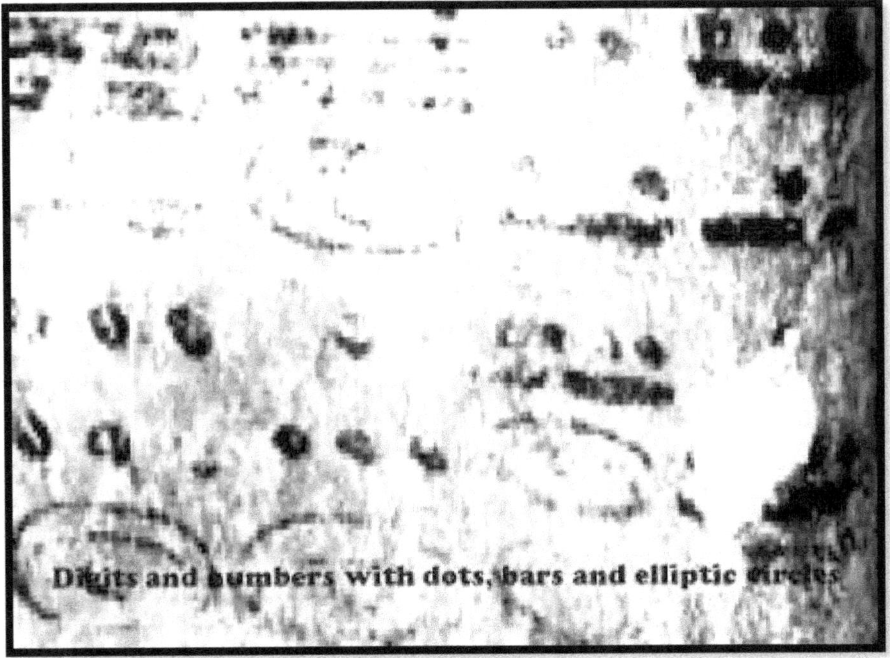

represent the 365-day solar calendar, the 584-day cycle of Venus and the 780-day cycle of Mars.

In the same wall the excavation revealed a portrait of the king, seated and wearing blue feathers, facing a figure in orange color holding a stylus. Markings near the latter figure's face call him "Younger brother Obsidian", who could be the younger son of the king, and the scribe who lived in the house.

The east wall was dominated with columns of numbers, represented by dots, bars and shell-like inscriptions, some of which tracked the moon or reconciled lunar phases with the solar calendar. Other tables of red numbers appear to be numerical corrections to make calculations more accurate.

Some calculations predicted astronomical events 7,000 years into the future. Contrary to some theories, there was no sign that the Mayan calendar ended abruptly in 2012.

The west wall shows paintings of three men wearing large, feathered black mitres, white loin clothes and medallions around their necks.

The major impressions of the different team members of the excavations was that Mayans were very attached in their daily lives and periodic ceremonies to the sky including moon, sun, planets and constellations.

Wall Paints showing men wearing medallions and feathered black mitres and white join clothes

Astronomer in his Observatory
Two rocket forms

CHAPTER V: MAYAN CITIES BUILT WITH STAR MAPS LAYOUT

He was enjoying the research about the Mayans and their civilization, a student with an age of fifteen years old discovered a lost Mayan's city, this was in 2016, but the story began earlier in 2012 when the end of the world theory was in its pick of curvature cause of the end of Mayan calendar. He searched in many manuscripts including mathematics, architecture, glyph writing, codex, time conception and astronomy, this last one was the stimulation for him to ask why the Mayas were interested in stars beside the idea of studying the constellations over the area where they reign for centuries. This smart teen called William Gadoury.

He stated to "Le Journal de Montreal" as he was Canadian: "I did not

Mayan sky watcher
Madrid Codex

understand why the Maya built their cities away from rivers, on marginal lands, and in the mountains, they must have had another reason, and as they worshiped the stars, the idea came to me to verify my hypothesis. I was really surprised and excited when I realized that the most brilliant stars of the constellations matched the largest Maya cities."

He started with location and identification of Mayan's major cities with large pyramids thanks to Google Earth tool, then he drew in transparent sheets or calks the constellations which the Mayas could see in their territories in Mexico, Guatemala, Honduras and El Salvador, between latitudes 13 degree and 22 degree north starting with circumpolar constellations of the northern hemisphere and finally, he compared the shape and angles of constellations with the geographical layout of the Mayan cities.

The starting point of his study was a sky watcher draw in page 33(this number remembers me about free masonry highest rankings especially when you know that this number is coming from angles between stars in constellations) of the Madrid codex one of famous Mayan writings.

With his point of view, the points around the man are stars and a group of those stars, related with dots-lines, means a constellation, so the Mayans were interested with constellations. There are 23 groups of points, so 23 constellations. After deep research, he noticed that the brightest stars of the constellations overlaid perfectly with the locations of the largest Maya cities. Those are some examples of his studies on constellations:

CASSIOPEIA:

One of studied constellation is Cassiopeia known as the seated Queen composed by five stars in the form of M or W, the brightest star which coincide with the main city of Mayapan which is the latest Maya capital is Alpha (Schedir).The similarity of forms between the constellation stars and the cities in the matter of angles is 96 % and the overall one is 97.2%.Those percentages reflects great similarity!

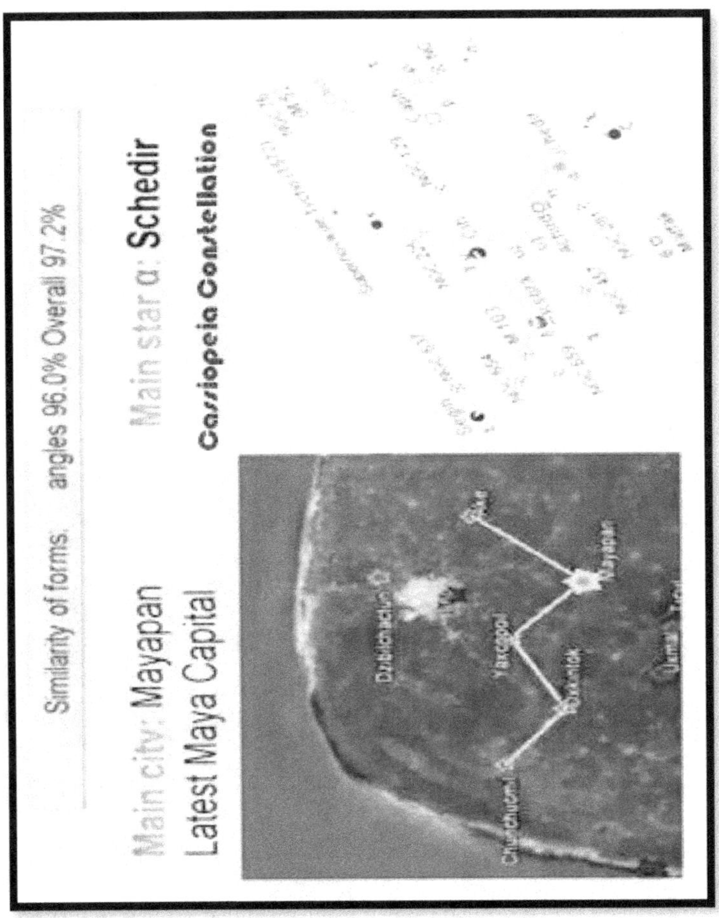

DRAGON:

The Dragon constellation known with its length and surround in part the Ursa Minor constellation it doesn't have a brilliant star, was studied too by this boy, the main star Alpha called Thuban which was the first polar star known with the emerging of the Maya civilization between 3000 and 2600 B.C .In the field we found the important city :El Tigre or Itzamkanac .The similarity of angles is 95.3 % and the overall similarity is 99.5 % .

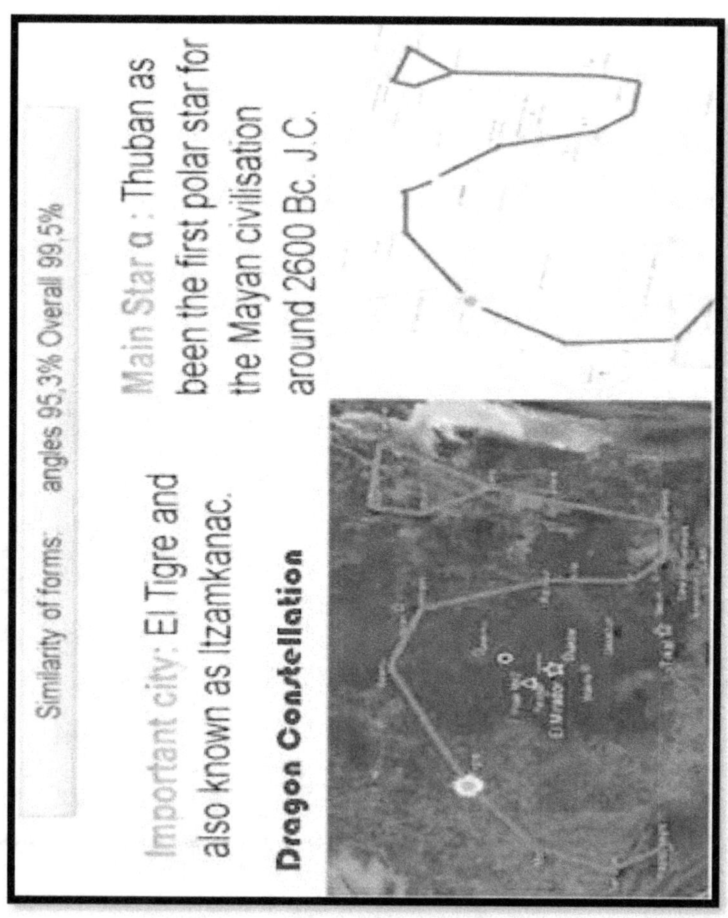

URSA MINOR:

The Ursa Minor constellation also called the Little Bear composed with seven stars the most brilliants are Alpha and Beta always in northern hemisphere .The main star Alpha is the Polaris A which is the third polar star for the Mayans, in the other side there's Becan as the important city.93.3 % is the angles similarity and 98.5% is the overall one.

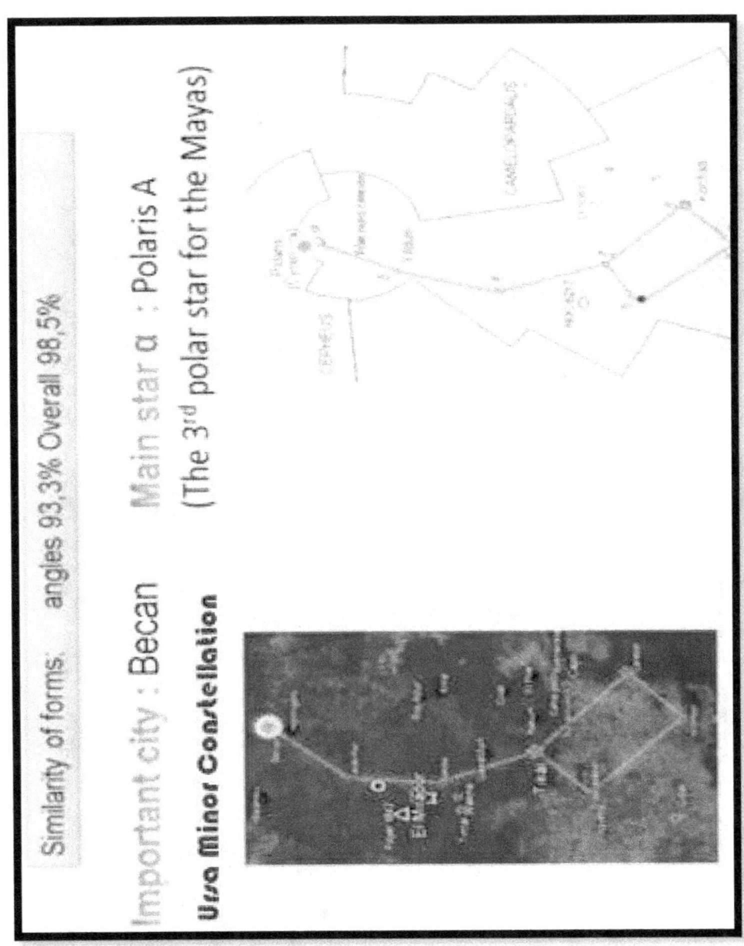

URSA MAJOR:

Ursa Major is one of tall constellations in northern hemisphere known under the name of Big Bear, the seven stars are also known as Big Dipper. The main star Alpha is Dubhe, on earth Tikal is the main city which is the largest city of the Mayan territory . The important star Alioth which is second brilliant in apparent magnitude of the constellation, in front we found Palenque which was distinguished by a particular architecture made by the king Pakal.The similarity of angles is 93% and the overall form similarity is 100%.Another conformity!

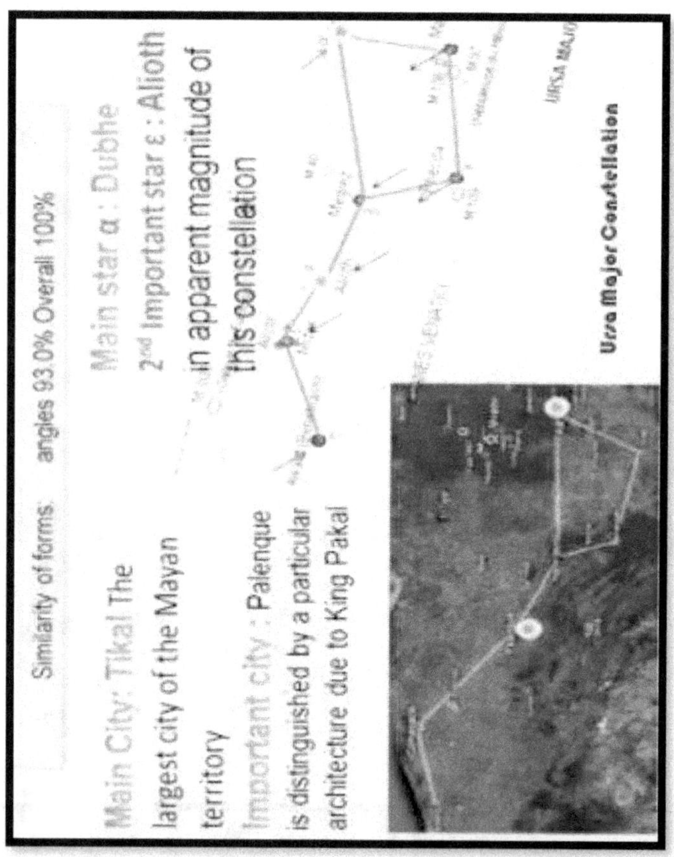

LYRA:

Lyra is a small constellation with six stars, 52nd in size occupying an area of 286 square degrees the most brilliant star is Vega which is the fifth brightest star in the night sky, after Sirius in Canis Major, Canopus in Carina, Arcturus in Boötes, and Alpha Centauri A in Centaurus constellation. Vega is also the second brightest star in the northern sky. The main city is Tikal which could reach one hundred thousand inhabitant in her pick.The forms similarity is 97.5% for both angles and overall.

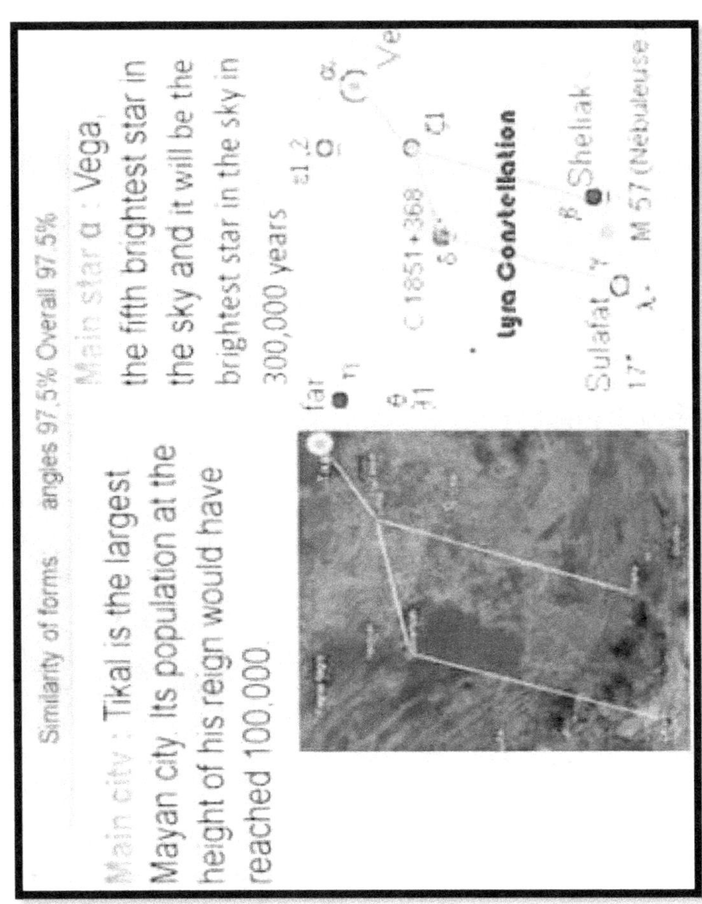

CYGNUS:

Cygnus is the 16th largest constellation in the night sky, occupying an area of 804 square degrees. It lies in the fourth quadrant of the northern hemisphere. Cygnus has 10 stars, the brightest star in the constellation is Deneb, Alpha Cygni, which is also the 19th brightest star in the sky .Deneb is an Arabic word means tail. The five stars that form the Northern Cross are Deneb (Alpha Cygni), Delta Cygni, Albireo (Beta Cygni), Gienah(wing in English) (Epsilon Cygni) and Sadr (chest in English)(Gamma Cygni) at the centre.

The main star Alpha is Deneb shining three hundred thousand times than our sun. In the other side the main city is ElBaul where the pyramids destroyed to make a way to a modern city. The angles similarity is 95.8% and overall is 98.6%.

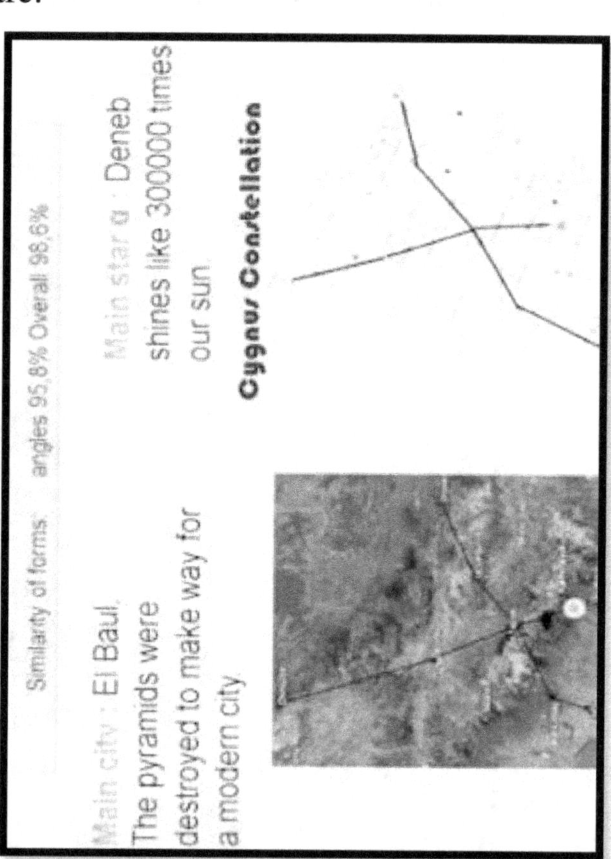

PEGASUS:

Pegasus constellation is one of the largest constellations in the sky; it's the seventh largest constellation in the sky, occupying an area of 1121 square degrees. It is located in the fourth quadrant of the northern hemisphere. The brightest star in the constellation is Enif, Epsilon Pegasi, with an apparent magnitude of 2.399. There is one meteor shower associated with Pegasus; the July Pegasids.The main stars are : Sirrah of Andromeda(Surrat Al-faras in Arabic which means the navel of the horse)in front the first main city :Caracol.The second main star, the third brightest in Pegasus is: Markab(the saddle of the horse).The main city is Kaminaljuyu.The similarity of angles is 96.7%.

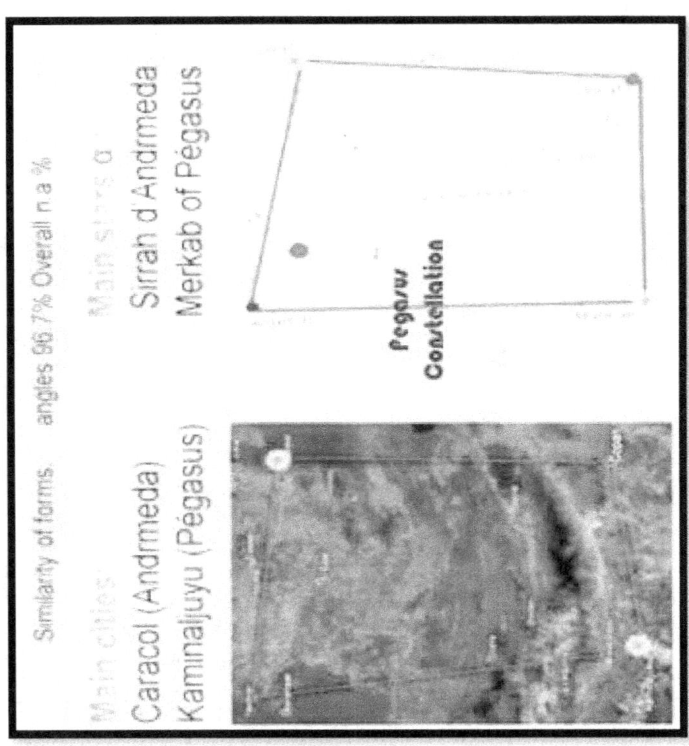

PLEIADES:

M45 Pleiades in the constellation Taurus, served with Mayan farmers .When the Pleiades appeared n the night sky, the Mayans knew it was time to plant and when they disappeared in the fall, it was time to harvest. The Pleiades, also known as the Seven Sisters. The brightest stars in the Pleiades cluster is Alcyone, the main city of Mayans is Cayo.The similarity of forms with angles 92.9% and overall 99.9%.

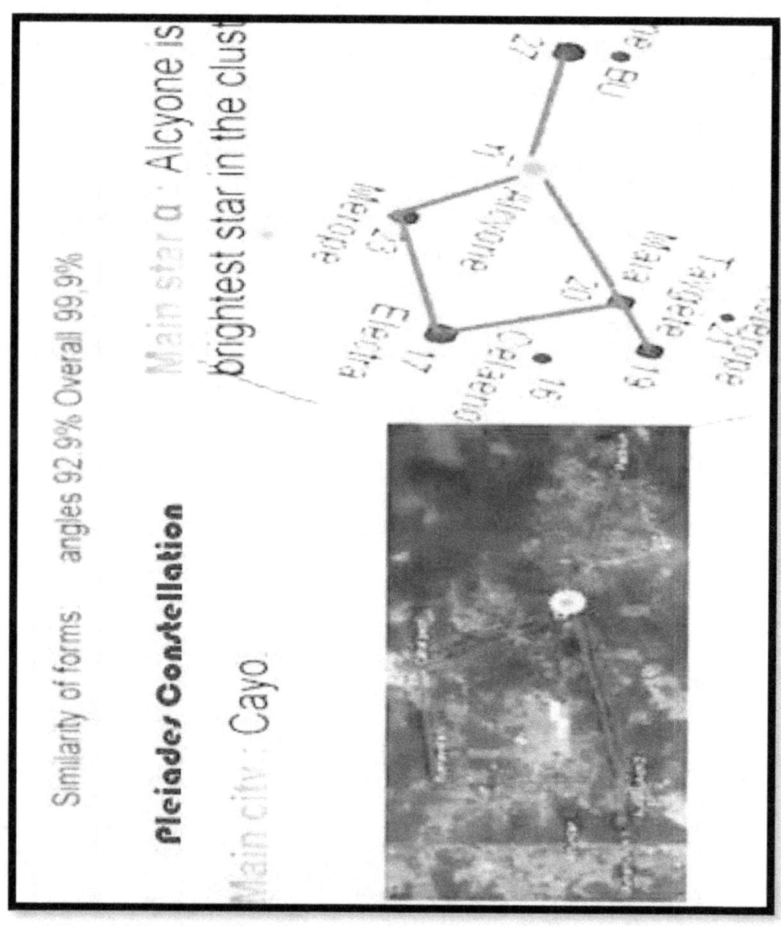

After an analyze with the G.I.S(Geographical Information System) of the Maya which is dedicated to localize and study the Mayans cities in Mexico, Guatemala and Belize, the 16 most important Mayans cities are represented in William's projects studies. From 149 second rank Mayans cities, 88 of them are too represented in his project with percentage of 60% which is a good percentage to give credibility to his method of studying.

From the 16 Major cities, 14 of them are represented by four brightest stars in some constellations like:
-The constellation of Ursa Major wth cities: Calakmul, Palenque, Piedras Negras and Tikal.

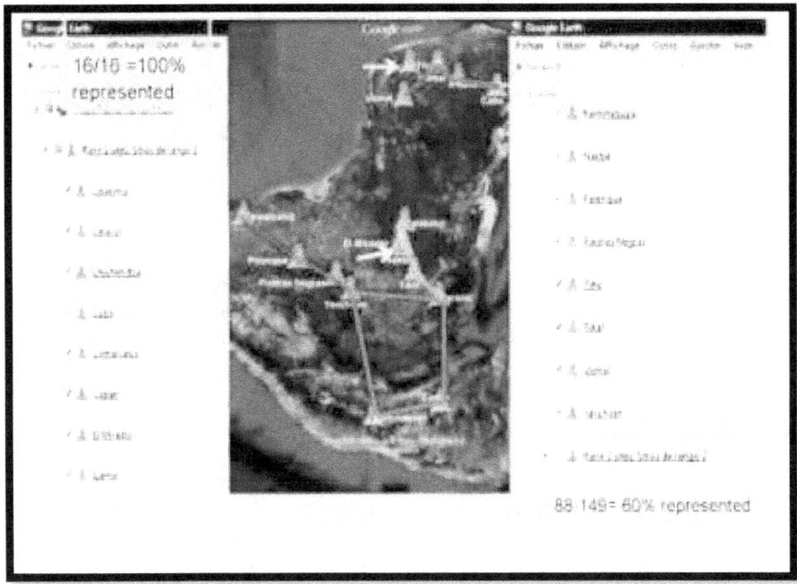

-The constellation of Ursa Minor with cities : Calakmul, Nekbe, Tikal and Caracol.
-Tthe constellation of Pegasus with cities: Caracol, Copan, Edzna and Kaminalyuju.

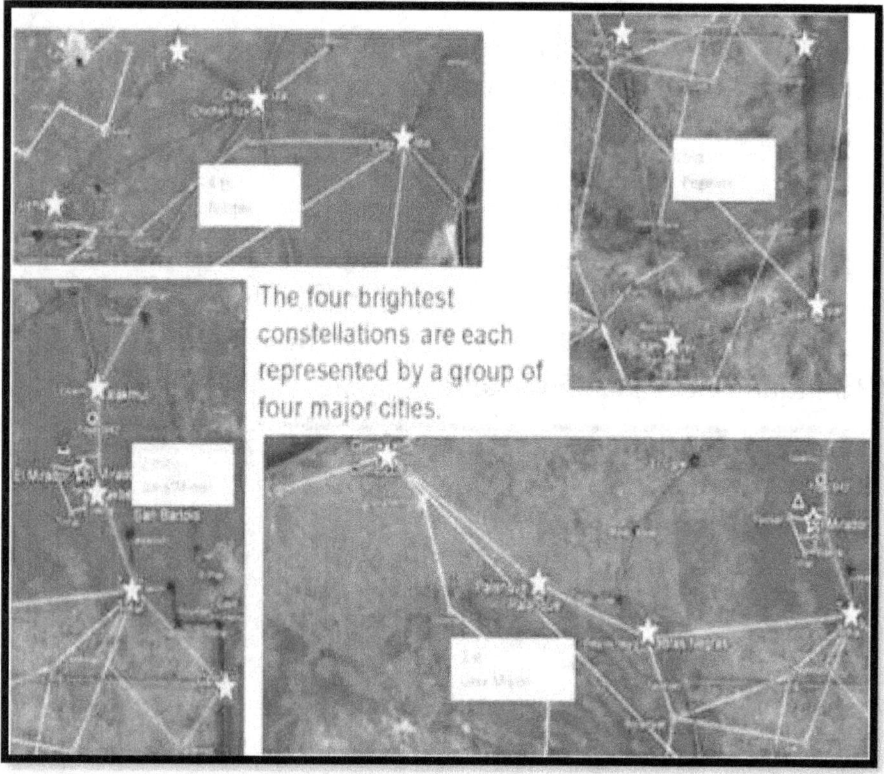

-The constellation of Bootes with cities: Uxmal, Izamal, Chichen Itza and Coba.

Also he proves with his study that the Mayans built their major cities according to the number of star for each city. As examples : city of Tikal with five stars, city of Chichen Itza with four stars, cities of Comalcalco and Coba with three stars and city of Copan with two stars.

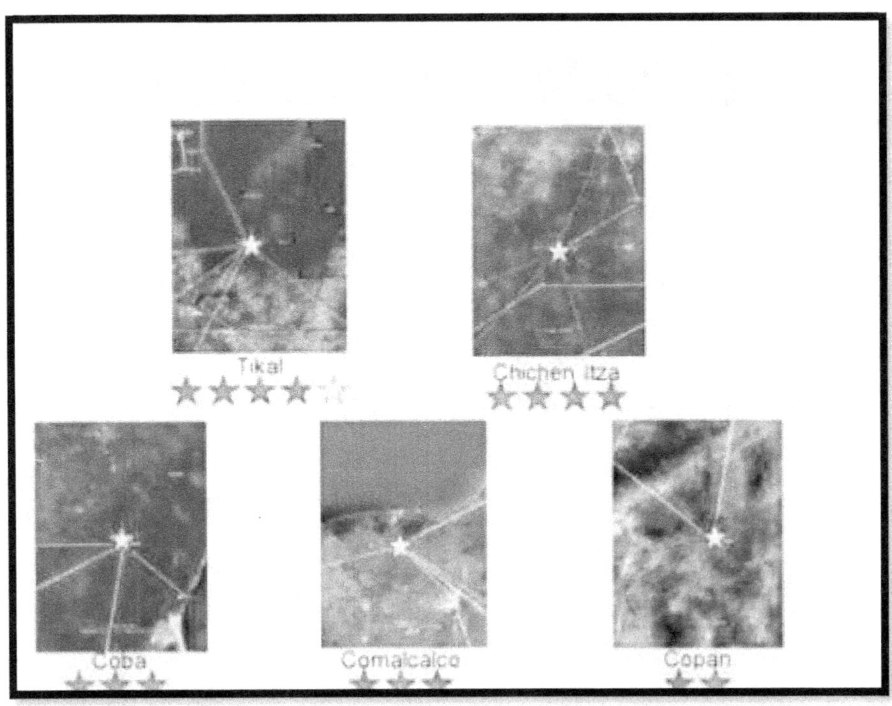

In 2014, he continued his studies and shows his work on Quebec science fair and IGARSS 2014, the 35 Canadian symposiums for remote sensing. His works achieve a Geoglyph Tree on Mayan territory of 325 thousand square kilometer dating from 4600 years ago. He approved 17 cities on territory by the projection of 22 constellations. A twenty-third constellation was missed based on his interpretations.

So he returned to the Mayan manuscripts Myths which is "The three hearthstones":

This myth is confirmed by Linda Schele who was a famous specialist in Maya civilization in university of Texas in Austin:

She claimed that the Mayans considered that the first creation was on August 13th, 3114 B.C.The three stones are identified as a jaguar throne stone, a serpent throne stone and a water fly throne stone. The act of setting the three stones is credited to Wak-Cahn-Ahaw or «Raised

Up Sky Lord" who has been identified as the Maize God.The three Hearthstones of creation correspond in the night sky to the stars Alnitak, Saiph and Rigel which appear in the belt of the constellation we refer to as Orion.

William determined the exact location of the three hearthstones on Maya territory which was in its center, something which can be considered strange and there was three main cities: Calakmul(in Rigel star), ElMirador/La Denta(in Alnitak star) and a missing main city between the

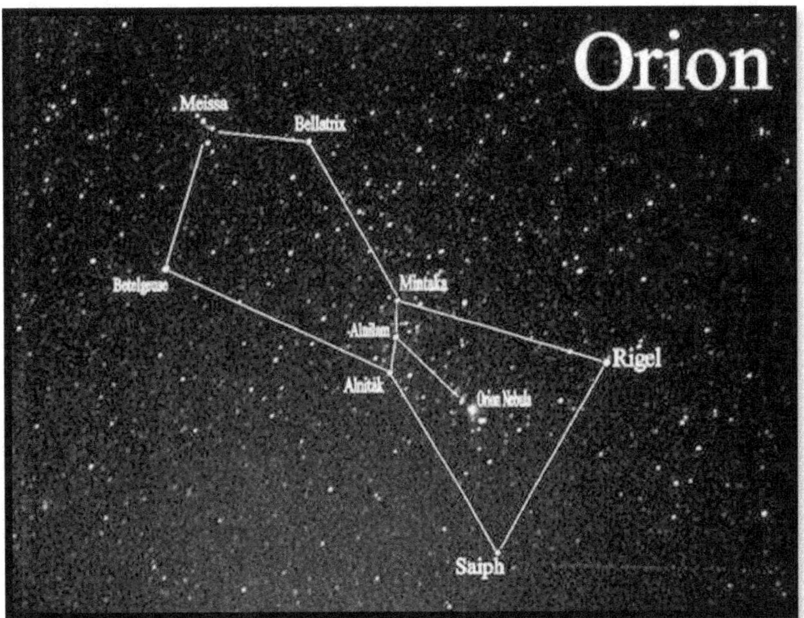

three in Seiph star.

The Canadian Space Agency helped him by making two satellite's photos of the area in September 12th and 14th, 2014 by RadarSat but the vegetation was too dense as he couldn't see anything via those images.

Then they gave him full time and access to search in this area and he took an image of a rectangular structure with two square kilometer which can be a platform supporting pyramids like the platform of ElMirador which supports the great pyramid La Denta .

LA DENIA

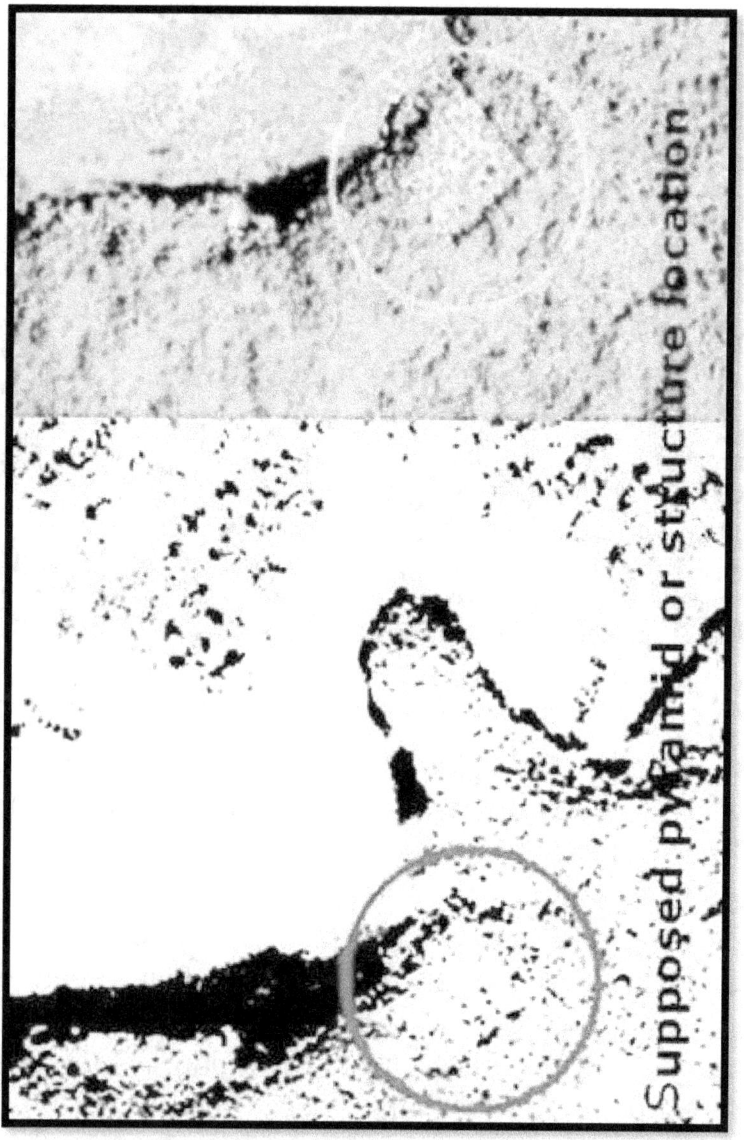

Supposed pyramid or structure location

Doctor Armand Larocque, a remote sensing specialist from the University of New Brunswick in Fredericton, believes it's a central Mayan pyramid, with several other 30 smaller nearby structures. He said too that the use of satellite

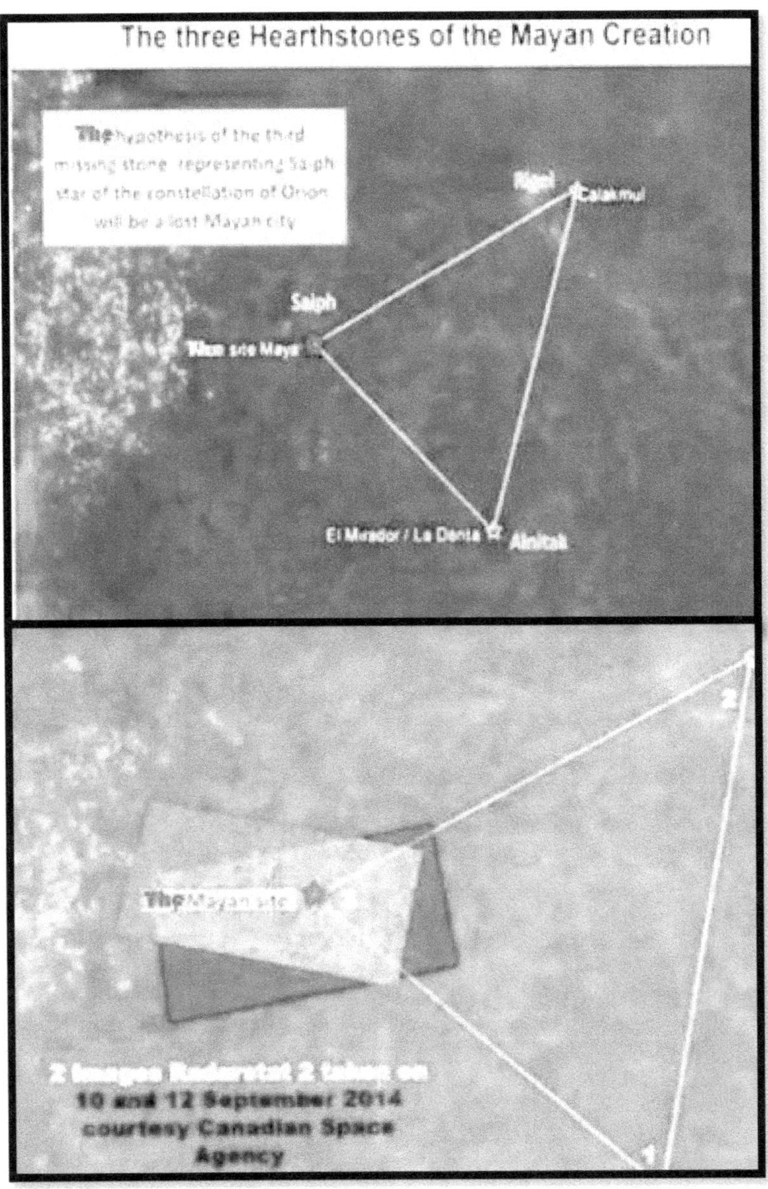

images, as well as the contribution of digital image processing, helped to confirm the possible existence of this forgotten city. "Geometric shapes, such as squares or rectangles, appeared in these images, forms that can hardly be attributed to natural phenomenon,"

Another hand help from the Japanese space Agency with Jaxa Satellite which gave him other radar images which presents great circular structure with 1.1 kilometer as diameter containing different structures mostly pyramids beside two great parallel lines which was interpreted by the National Institute of Historical Anthropology as a great enter to this city.

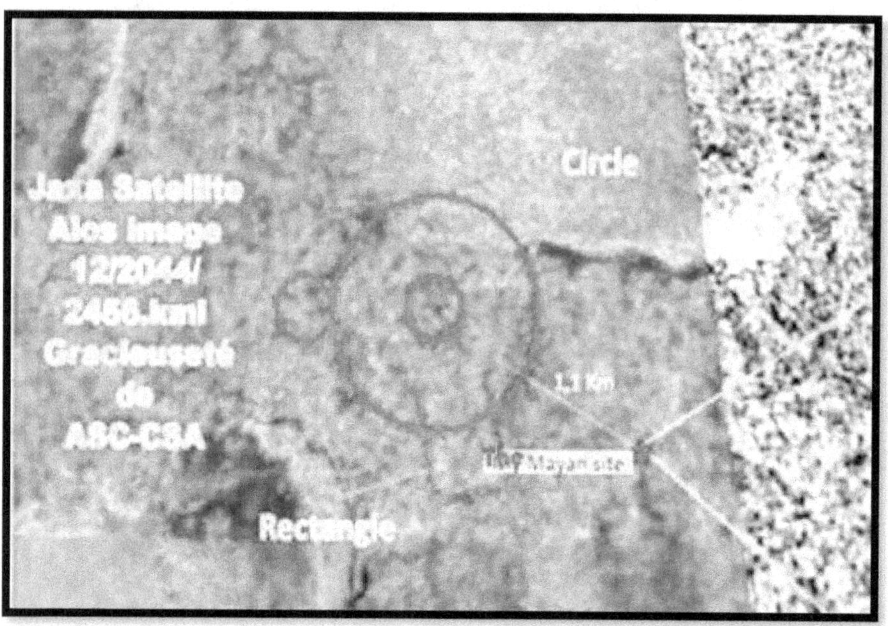

Another image from an online Geographical information system taken in January 17, 2006 showing this area of four by four kilometer equivalent to 16 square kilometer which is the third Mayan Hearthstone. It was taken days after a great fire in Yucatan Peninsula. Structures or pyramids emerge separated by Mayan artificial ways. William gave the name of KAACHI, the fire mouth to his city.

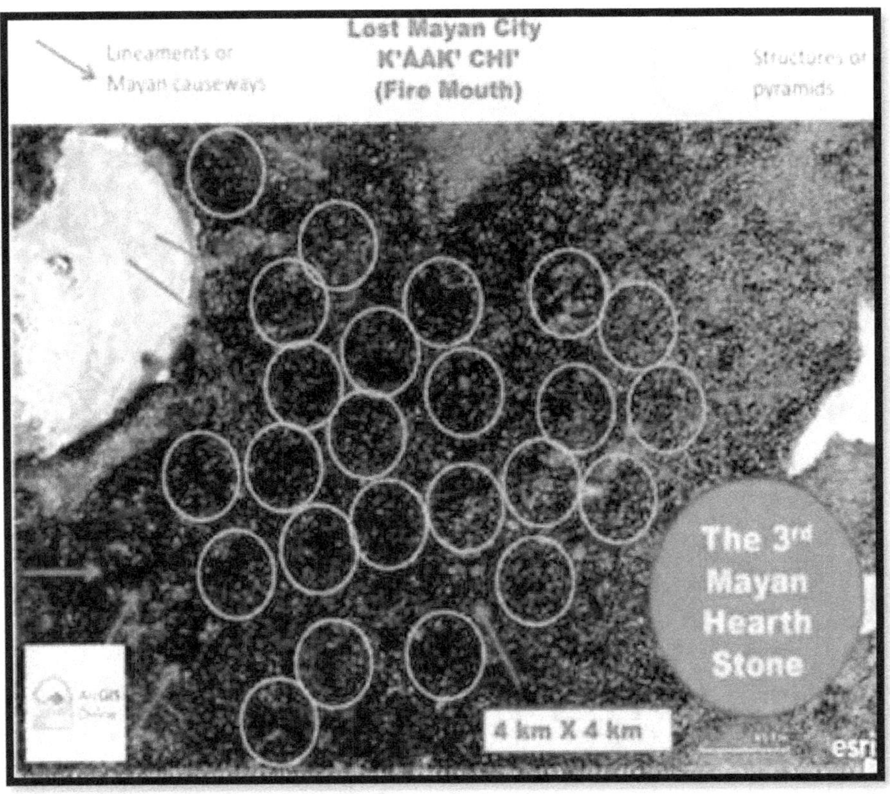

In this same area, there are two accesses from east and west for the structures represented here with white squares, The Green and red circles had two accesses too from south and north. The surprise here is there's alignment with the sunrise and sunset at the two equinoxes. Daniel de Lisle from the Canadian Space Agency who was involved in Gadoury's researches said that he was fascinated by the

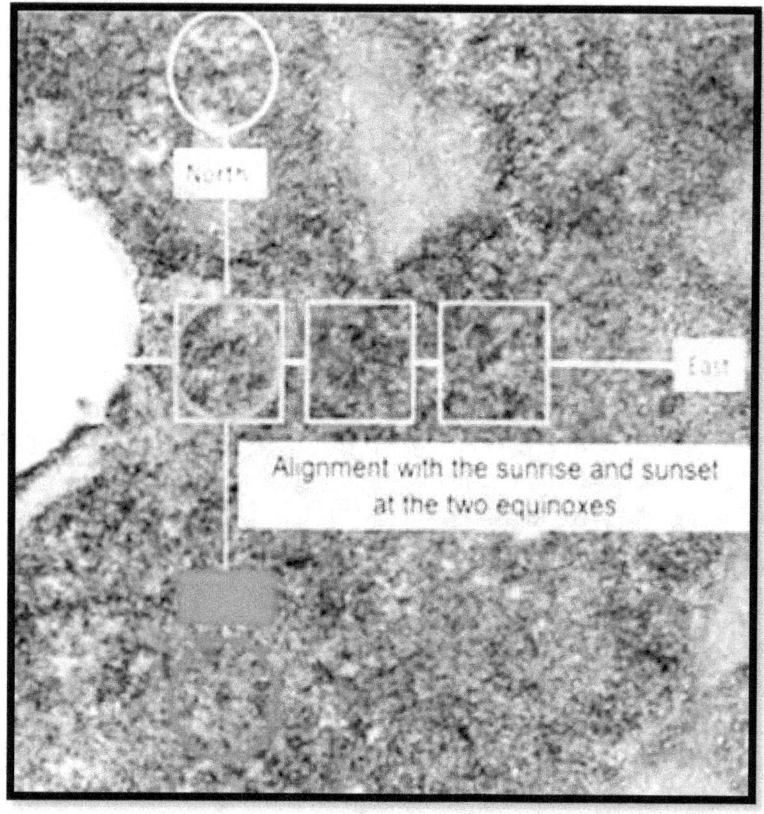

quality of those researches especially thinking about making a connection between stars and lost Mayan cities. He describes it as very exceptional. He declared to a Newspaper that "There are linear features that would suggest there is something underneath that big canopy," adding that "There are enough items to suggest it could be a man-made structure.

KAACHI the supposed new discovered city can be considered as one of greatest city in the Maya civilization.

Comparaison with the five (5) most important cities			
Cities	Area Km²	Area Km²	Highest m
Rank 1	City center	territory	structure
Tikal	16	120	65
El Mirador	26	100-150	70
Calakmul	20	70	45
Coba	3,5	70	42
Chichen Itza	6	25	24
K'ÁAK' CHI'	18	80-120	86

K'ÁAK' CHI' is one the four most important cities

Critics: Dr. Armand Larocque statements were not convincing for the academic society. Doctor David Stuart an Anthropologist in Mesoamerica history from university of Texas at Austin wrote in his Facebook account: "This

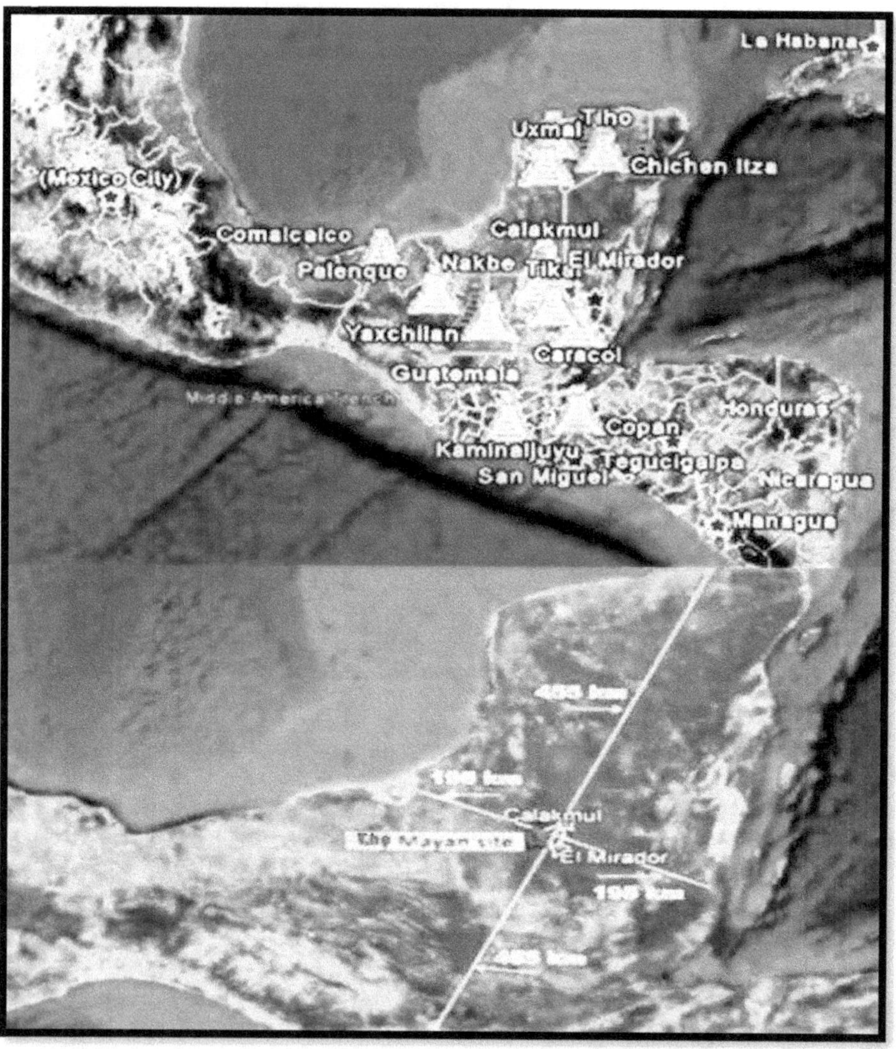

current news story of an ancient Maya city being discovered is false, the whole thing is a mess – a terrible example of junk science hitting the internet in free-fall. The ancient Maya didn't plot their ancient cities according to constellations. Seeing such patterns is a rorschach process, **since sites are everywhere, and so are stars**. The square feature that was found on Google Earth is indeed man-made, but it's an old fallow cornfield, or milpa.

My answer to Doctor Stuart when he said: "…**since sites are everywhere, and so are stars**…".Sure sites are everywhere so what are you waiting for to discover them and stars he means constellations of stars are everywhere, yes they are but for thousands of years, they disappears and reappears year after year in the same astronomical positions each depends on the human observer on earth, east, west, south and north, so they are periodic visitors for thousands of years and this is increase the potential of using these stars as references for so many ancient civilizations around the earth like the Sumerians, the Pharaohs as examples, the Mayans are One of them. The same for planets which appears too every year and year after year in the same places, Venus and Mars are references for Mayan calendars. It's clear that Doctor David Stuart has not even little knowledge about constellations and never thought to including this factor on his researches, he concentrated just on the ground.

Doctor Stuart forgot that the Mayas Codices are full of astronomical tables and zodiacs.

Another anthropologist from University of California, San Diego, Doctor Geoffrey Braswell explained that the zones shown in satellite images are "places that are well known to archaeologists who work in the area" and that the images are "not of Maya pyramids" but can be active marijuana fields.

Doctor Thomas Garrison, an anthropologist at University of South Carolina Dornsife and an expert in remote sensing, said that these objects are relic corn fields , the rectilinear nature of the feature and the secondary vegetation growing back within it are clear signs of a relic milpa. I'd guess it's been fallow for more than a decade. This is obvious to anyone that has spent meaningful time in the Maya lowlands.

He showed this image from Guatemala as an example to discredit Gadoury's interpretation, but it's clear that this abandoned field is curved but the young man images shows the contrary uncurved area, a tall structure overall.

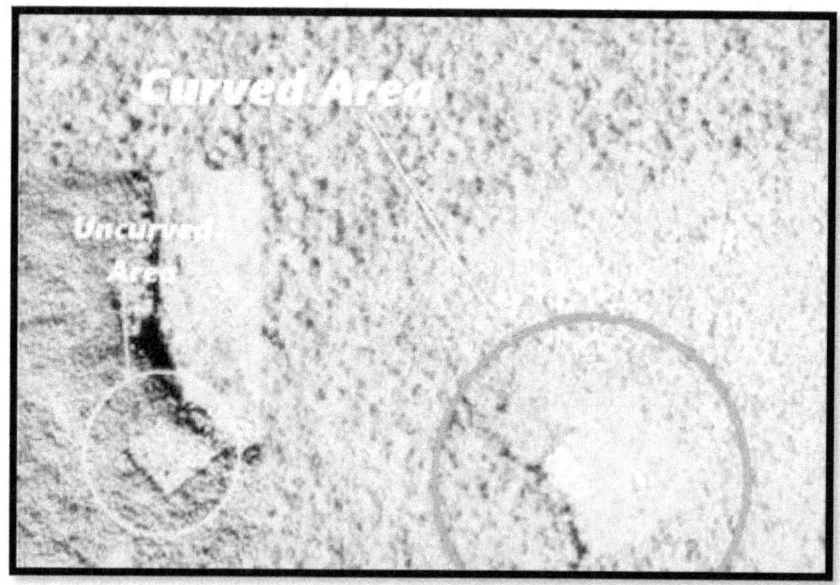

Doctor Ivan Šprajc from the Institute of Anthropological and Spatial Studies in Slovenia, said that the Maya were very good astronomers, and that they were interested in certain stars and celestial objects, but he's skeptical that these charts can be used to reveal the location of Maya sites. Very few Maya constellations have been identified, and **even in these cases we do not know how many and which stars exactly composed each constellation**.

As an automatic answer to this statement, William Gadoury did the work with 23 constellations and find 117 cities. All these declarations are general and marginalized. Even we consider that this area is for marijuana and corn, it's strong possible that under it located a Mayan city.

CHAPTER VI: HOW MAYAS WERE DISCOVERED

In 1839, an expedition left The United States of America in seeks of the lost Mayas city, Copan in Yucatan Peninsula. This expedition was commanded by two adventurers the first one was John LIyod Stephens a Lawyer and Author, the second was Frederick Catherwood a British Artist. They found it buried under the rain forest and uncovered massive pyramids, stairways, platforms and buildings. This discovering confirmed the rumors about this mysterious city and it was a first step to uncovering the entire ancient civilization.

At their arrival, each one took a specific mission for him, John LIyod Stephens write about their diaries, what they saw, discovered and all the chronicles. Frederick Catherwood painted the pyramids, temples, nature, Mayas and all what he could catch.

After exploring Copan, they made a long travel to the two ancient cities Palenque and Uxmal.This adventure was cut shortly by the sick of Frederick Catherwood with malaria, they were obliged to return to New York.

Stephens wrote: "…On our first expedition, we took a boat from New York and arrived in Belize city. From there we traveled south toward the first ruin, Quirigua.Just south of Querigua, revolutionary soldiers imprisoned us in an abandoned church for a day. After our imprisonment, we stayed in the wonderful ruin, named Copan.There we dealt with a man who claimed to own the idols, and we bought Copan's ruins for 50 $.After Copan, we decided to search for the ruins of Palenque. We traveled through Guatemala City and Comitan before heading north to Palenque.While in Palenque, we stayed in the wonderful ruins, but suffered from suffocating heat and rains as well as snakes, insects and scorpions.Catherwood suffered from malaria.We left Palenque, boarded a ship and traveled up the western peninsula coast to the wonderful city of Uxmal. The effects of malaria made Catherwood delirious and he

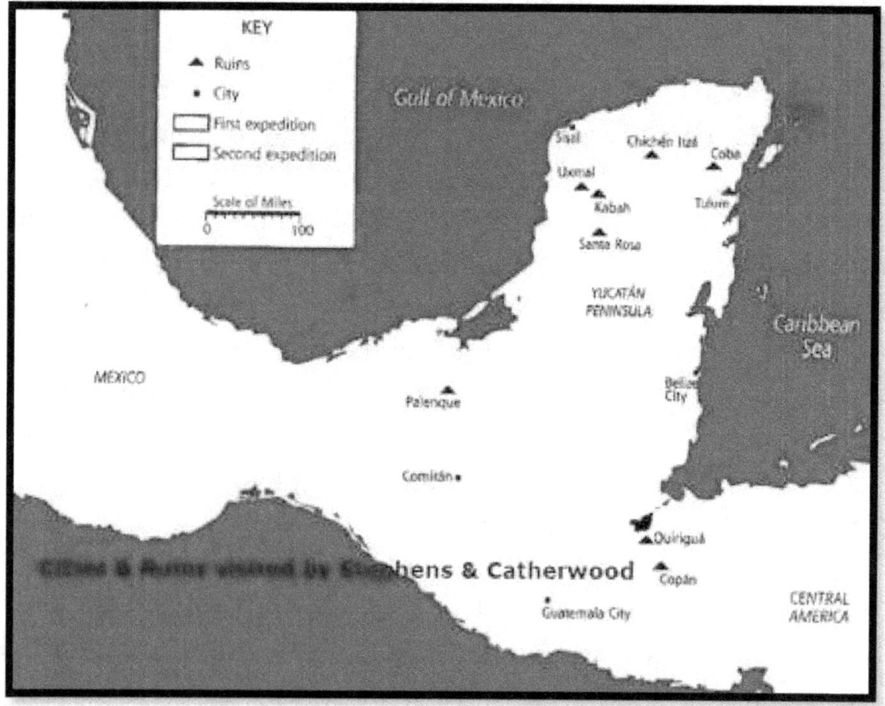

collapsed. We decided to head home to New York so that he could recover."

Their second travel to Central America was concluded by publishing a second book by John LIyod Stephens but he didn't mention well the works of Frederick Catherwood something which influence their future decision to not travel together again.

Stephens wrote: "We arrived on our second expedition in Sisal. We set out to finish our work in Uxmal, which took six weeks. Then we headed south to a ruin unknown outside of Mexico called Kabah. We spent several weeks at Kabah exploring, taking notes, and drawing pictures. Next, we traveled to Santa Rosa.

We arrived to Santa Rosa, where the only food was iguana. Then, we traveled north to the most famous ruins of The Yucatan, Chichen Itza, where an ancient road led us to great sites. From Chichen Itza, we went north to the coast, boarded a boat and traveled to Cozumel island for our next exploration. We crossed the waterway from Cozumel island and explored Cuba and Tulum. After that, Catherwood became sick again, so we headed back home- never to return together.

The books of Stephens opened the doors to curiosity to make more expeditions and be interested in Mayas and Central American civilizations.

CHAPTER VII: MAYAS' ORIGINS

Mayans are the indigenous people of Central America characterized today in the countries of Mexico, Guatemala, Belize, El Salvador, Honduras, Nicaragua and Costa Rica. Not all the Mexico was a Mayans homeland just the central area extended with the geographical south towards the rest of the countries except parts of Honduras. The main cities inhabited by the Mayans are: Teotihuacan and especially in the Yucatan Peninsula covering the cities: Palenque, Tikal, Chichen Itza, Mayapan, Uxmal and Copan.

The Mayans were shorts; the average height of a man was five feet and for women was four feet eight inches tall. Their hair was deep black and so many of them colored their bodies with black, red and blue.

They put tattoos; they appreciate the crossed eyes and tied objects from their infants' foreheads to make their eyes crossing.

The date of creation of the Mayans which is also the date of the first calendar used by them was 3114 year Before Christ. The pre-classic period was between 1000 B.C and 250; during this time, the Mayas came in contact and then borrowed by the Olmec Indians. The early Mayan settlements were fishing villages along the Pacific Ocean and Caribbean Sea. Then they moved inland and became farmers in northern Guatemala. After four hundred years, they began to build large pyramids.

After a complete eight centuries and half, at the year 250, the classic Mayan period started with building of great cities, Tikal was the largest, its spread over 50 square miles and population over 100 thousand. The central plaza in Tikal measures 250 by 400 feet. Two of the eight temples of Tikal face each other across the great plaza. The temple of the great jaguar and its pyramid rise over 150 feet. Archeologists discovered a skeleton of a ruler called Double Comb inside one of the pyramids beside him Jade, pearl and Shell Jewelry. Copan is the second largest Mayan city, it contains five main plazas.

The most famous ruin in Copan is the great staircase, it's 30 feet wide and has 63 steps. Picture writing covers each steps. Copan has a perfect ball court for playing. The Chichen Itza city's ruins has several plazas, pyramids temples and ball courts. The great pyramid can be seen miles away. The large observatory tower ElCaracol used for astronomy stand up with good state.Chichen Itza is the location of the well of sacrifice where the sacrificed men for their gods are thrown. The writings, the arts, architecture, astronomy and mathematics flourished with a sustainable and stable life.

Between the years 800 and 850, the Mayas begin to leave their cities, which was and still the great mystery, they scattered through the countryside, the records used by them are missed no book papers or Stelae. The Stelae is a large upright stone monuments which has carving of a man in one side and hieroglyphs in the other side. The Stelae beside Books Papers are used to commemorate daily normal and religious events ceremonies, information about rules, priests, governments and communities. The information was recorded at least every twenty years.

Archeologists and Historians supposed so many theories about the cause of this sudden collapse. No one knows what's really happened because nothing of their writings describe this period of history. Some people believe that a natural disaster like hurricane, earthquake, volcano were behind the deserting of the Mayas cities. An epidemic of a disease occurred like yellow fever something makes them leaving so quickly that none could have time to record what was happening.

Others, thinks that Mayans left due to a famine or potential famine as the Mayan's agriculture system caused the exhausting of the soil as they burned the remains of crops a method known as slash and burn the oldest one in agriculture history, the ashes are burned to be fertilizers for the soil, it takes one to three years to be a fertilizer, the Mayas didn't give this time to burned ashes, they burn until the land became clear then they replanting again, this destroy the fertile quality of soil something which can't yet feeding the growing population.

They didn't developed new ways of farming and agriculture; they used always the planting stick and never using a plow that's why the crops sizes were limited. Animals were not parts of transportation process nor the wheeled vehicles. The only way to transport the food from the planting fields was the manpower including women. This maybe limited the potential to plants long distance and big surface farms. If this is true the Mayans could leave their cities to search to other opportunities.

Another theory suggest that peasants revolted against the rulers as they work very hard in farming and transportation of those farms and the counterpart was not satisfactory, the peasants left cities and rulers hadn't workers who served them, no food and they leave too. The last suggestion is the war between them and invaders, a short one which don't give time to write.

All these theories even if they are true; missed one thing which is the skeleton remains of Mayas where are they if all these scenarios are really occurred or even one of them and those remains should be dispersed in many places inside those cities and not be buried in a regular form like the collective tombs in caves as example and even if we take this hypothesis it can't be done because all were disappeared without trace. The human skeleton stays for thousands of years it can't be dissolute like liquid and be absorbed by soil! No one of archeologists found skeleton remains of this period of Mayan history!

It could be a collective abduction from extraterrestrials huge mother-ships which they considered by them, gods? Like these ones:

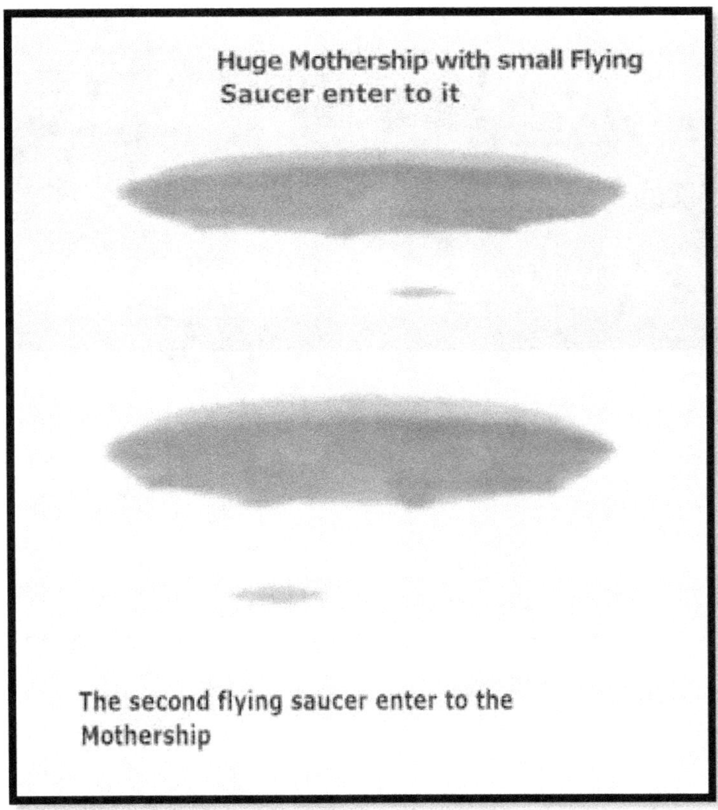

Huge Mothership in Alps Mountains
Flying Saucer in
Sea

At the year 900 and after the mysterious collective disappearance of Mayans from their cities, the post-classic period of Mayas history began and the Chichen Itza city became the most important city. Mayans daily activities like art, writings, games, etc goes well. After 300 years, Mayapan took the place of Chichen Itza as the main important city of the Mayas. In 1440, a long raged war occurred between the Mayas leaders until the Spanish conquest in the year 1527.Mayas resists to Spanish for long time but at the end the colonizer won, killed tens of thousands and destroyed thousands of writings which can be very useful today to know more about the Mayas. Today Mayas lives with millions and still conserve their ancestor's rituals.

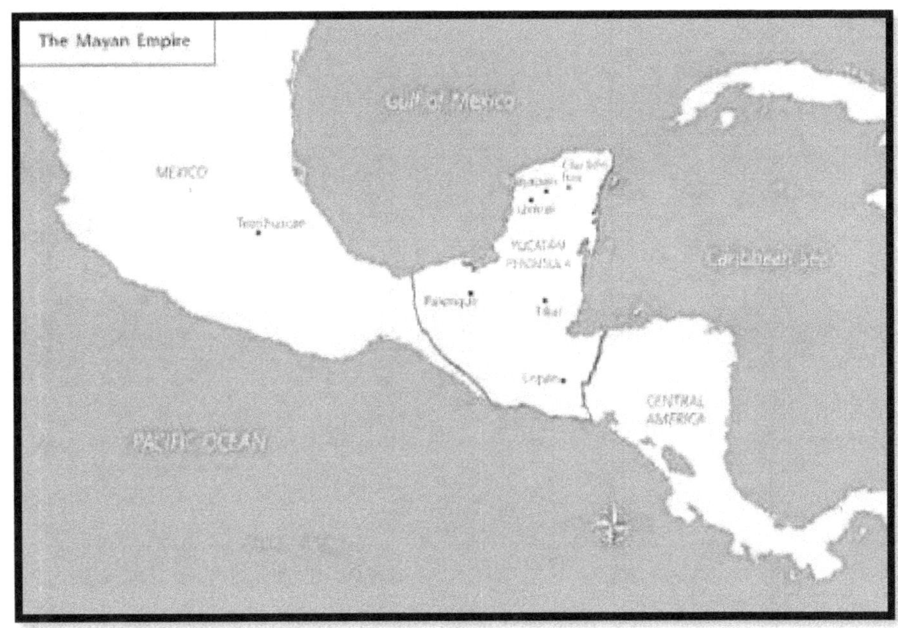

CHAPTER VIII: MATHEMATICS

Mayans developed a mathematic system which was considered more advanced than other civilizations like the Greeks. The use of the zero by the Mayans was very early in comparison with other ancient tribes. It was a very important invention. They used a picture of a shell to equal zero. The dot is used as equal to one, the bar equal to five and they use a base of twenty instead our use the ten as base. Smaller numbers were written horizontally while larger numbers were written vertically. For each position going up, the column represented a multiplication by 20.

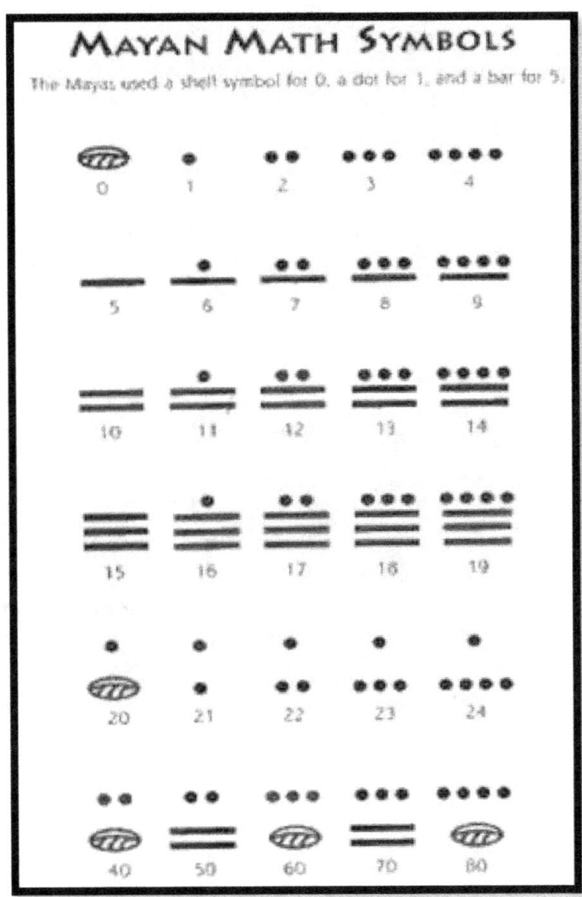

CHAPTER VIIII: GREYS ARTIFACTS

The museum of Xiutetelco in Puebla, Mexico announced the discovery of a statue which depicts a "grey alien figure". The statue was uncovered by municipal workers in a village called Maxtaloya located in the city of Tepeyahualco. After careful cleaning and manipulation, archeologists identified its composition from obsidian, turquoise, jade and shell.

The humanoid figuration is with large oval dark eyes surrounded by a snake and holding or sitting on a human Maya's head, the obvious reasons it is described as "Grey alien" are the shaping of its large head and the oval eyes that are known for this kind of extraterrestrials. Some specialists, thinks that the town was covered by a volcano eruption in ancient times. Another strange artifact was found which has the form of Reptilian and Grey in the same time.

The director of the museum, Mr Rafael Julián Montiel, declared that more researches for artifacts in the area will be done because of its close proximity to Cantona (La casa del sol) which is a one of the largest pre-Columbian cities discovered in Mesoamerica and contains Mayans architectures like Temples, Ballcourts and Plazas and have barely been excavated. Archeologists estimated that 90 to 99% of Cantona still not discovered. They urge to make the excavations fear of nearby volcanic eruptions. With this Grey humanoid statue, another several thousands of

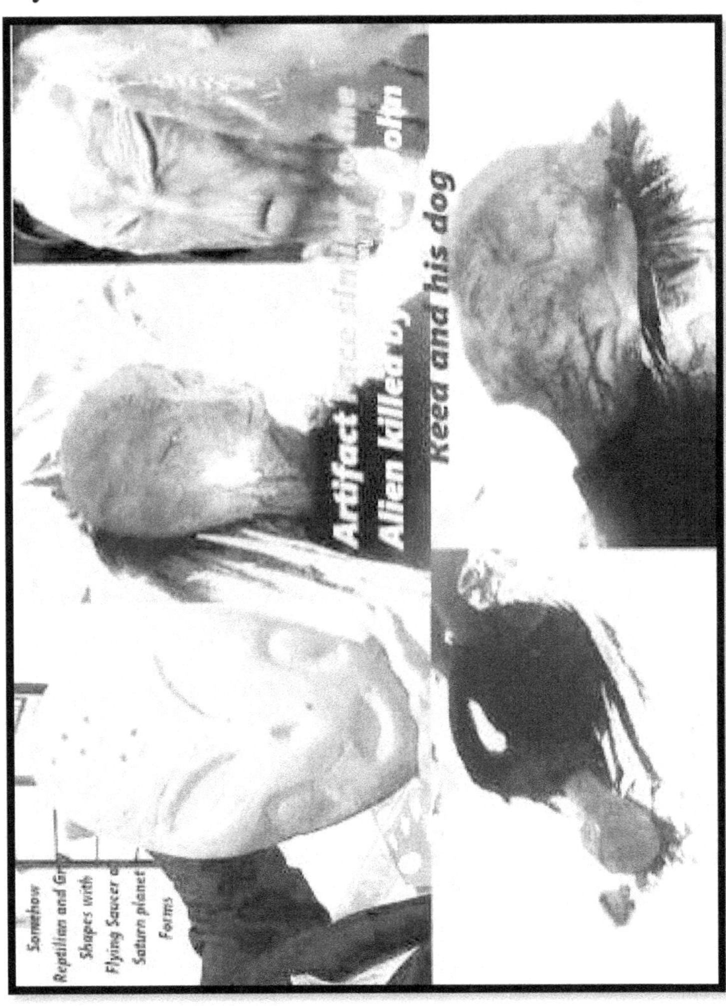

artifacts were also depicted in this village and all were taken to the Xiutetelco museum.

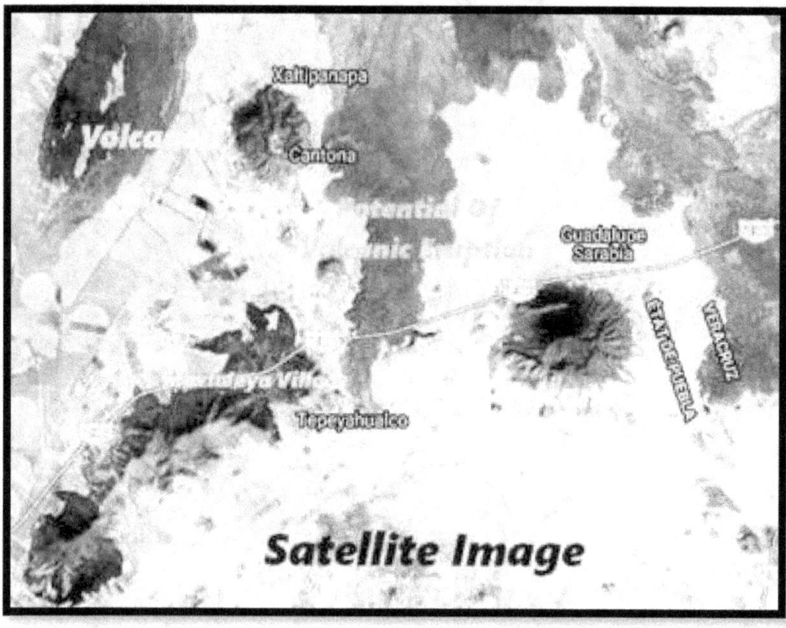

Like the wall arts, Stelae and codices, this statue reflects a certain Mayan mythology and culture. The snake or serpent mostly represents the Quetzalcoatl, the feathered serpent deity and god of wind and learning. Serpents were also used to symbolize the underworld and a means of transportation or transfer from the underworld to the heaven, which could be explained:

The Mayan head is a man in dying process transported by the serpent to the heavens which is represented by the Grey aliens who visited the Mayas several times. From the border region between Veracruz and Puebla states a local famous journalist called Javier Lopez Diaz working for the radio Cinco Radio claimed in his twitter account a discovering of Mayas artifacts:

Twit 1 :

El explorador poblano Jac encontró en los limites de Puebla y Veracruz piedras de jade grabadas con hombres y extraterrestres (1/2) pic.twitter.com/Y6yMBW9v3A

The Puebla explorer Jac found jade stones engraved with extraterrestrial figures in the borders of Puebla and Veracruz.

Cinco Radio (@JavierLopezDiaz) 24 de abril de 2017

Twit 2:

Se muestra la mezcla de dos culturas. Las piedras podrian significar el contacto alienigena con los mayas o aztecas (2/2) pic.twitter.com/cS5tOHvzFU

The mixture of two cultures is shown. The stones could signify alien contact with the Mayans or Aztecs

Cinco Radio (@JavierLopezDiaz) 24 de abril de 2017

The Puebla Explorer is known as Jac Detector(Twitter: @jac_cetector) his true name is Jose Aguayo.He's one of the team of researchers whose investigated a cave in which they found stones with relief scenes depicting aliens, as well as spaceships of aliens. The search for mysterious caves began with legends in which it was said that in ancient times alien beings arrived on a spaceship and ancient people captured it on stone bas-reliefs that were made in a system of three caves located somewhere between the cities of Veracruz and Puebla in Mexico.

Jac Detector
Members of the Research
Team

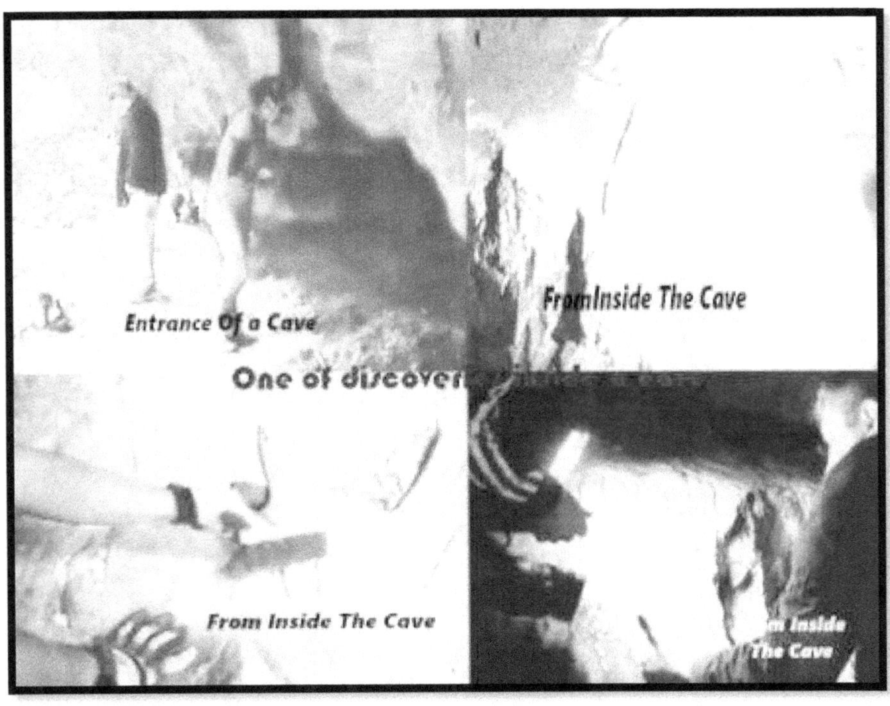
Entrance Of a Cave
FromInside The Cave
One of discover
From Inside The Cave
m Inside
The Cave

Interpretation for some Artifacts:

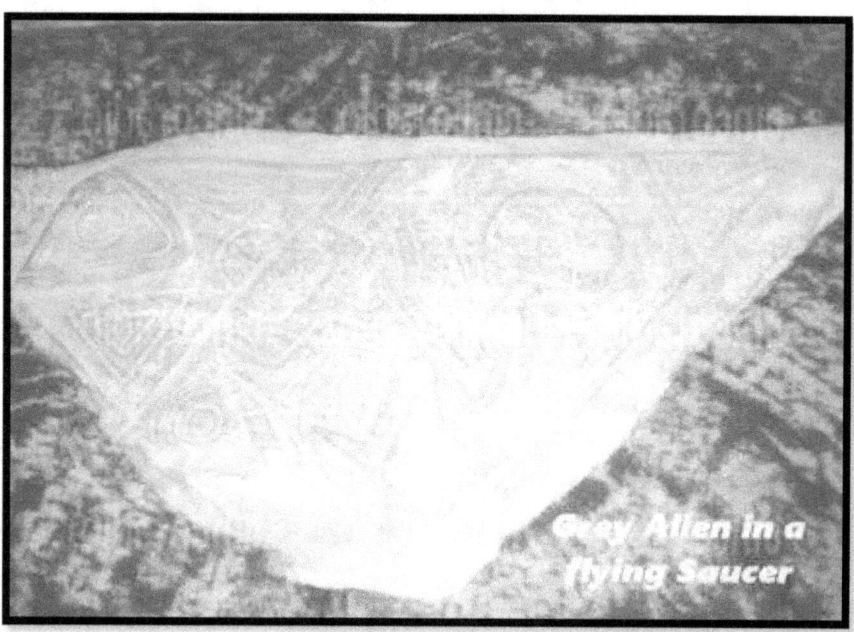

A Grey Alien with a hand of three fingers piloting a triangular flying saucer which was reported in the eighties of the twentieth century and still in our days for many sightings. He or she has a bands around his head, his arm, forearm and his leg.

The Artist who did this work want to explain that the flying saucer has the form of a triangle. You have the feeling that he knows so many details inside this Spaceship and this can give the idea that the Mayas possibly saw everything within it or why not, travelled with it..Light bulbs or flashes existing in the corners of this flying saucer. In the Artifact, one is missing, in the corner where the Grey Alien is sitting.

The Grey Alien in this Artifact has great resemblance with these Greys below, they are small, they have pink-white color and not Grays as thought, three toes are clear so automatically three fingers despite they are not clear in these photos:

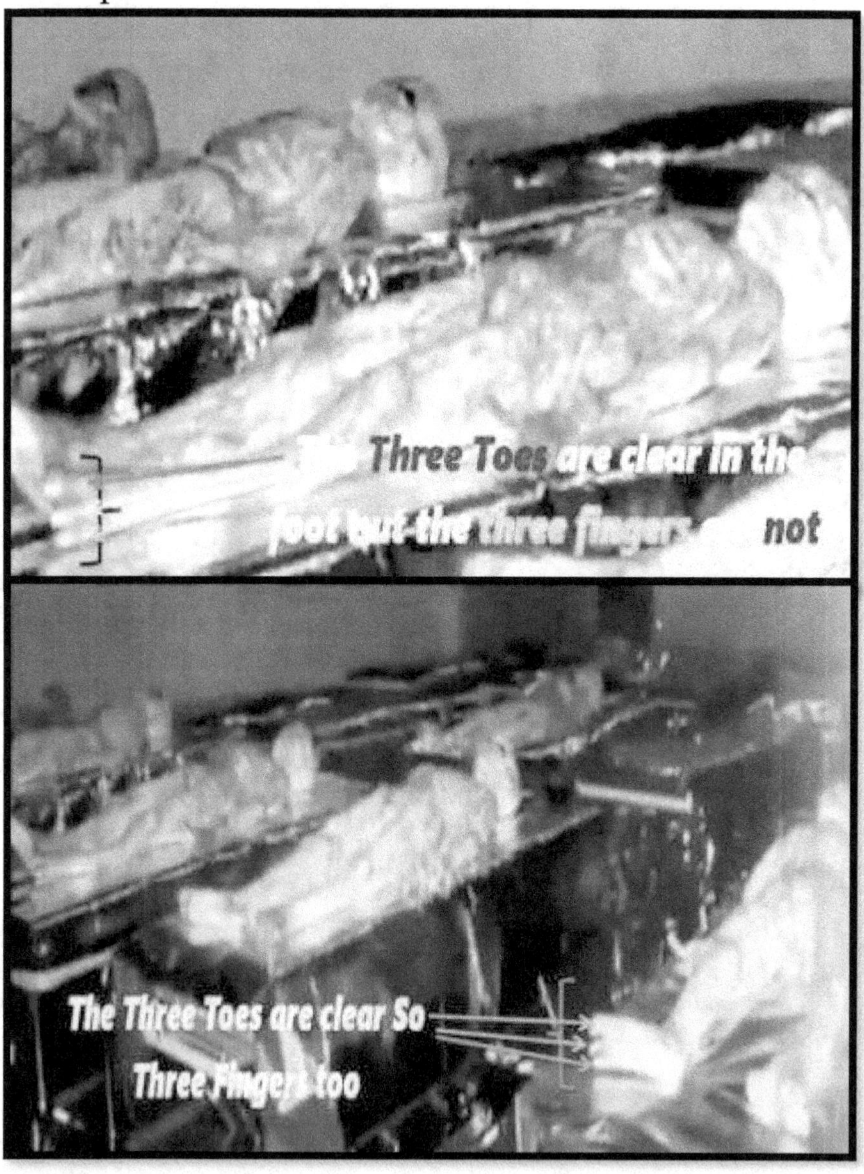

In this image, there's an exchange of gifts between a Mayan in his knee and a Grey Sitting over one knee means great level of respect from the Mayan to the Alien, it means a moral authority by the Aliens over the Mayas. This moral authority can be explained by the coming of Aliens from space with their flying saucers and Mayas worshipped planets and stars so this deep respect is coming

from this perspective and this can explains too two other things: the frequency of visiting of Greys to the Mays was very high and the cause of building pyramids and observatories are results of these repetitive visits. The flying saucer is round and it shows some kind of festivities with an open door or window making lights dancing show. The Mayan's gift is appearing as a fruit but the one of Grey sure unknown. The Grey Alien is the same as in the first Artifact.

In this Artifact, we have two Grey Aliens: the first one is over a box and the second is taller, his head is bigger and

taller, the first one is mostly a kid and the second is a father or a mother talking to a Mayan Priest.A female or a male Mayan is inside a box.

The two Aliens one has a kind of a ball and every ball has a geometric form.These balls can be the balls used in Ballcourts games.The geometric forms: square, triangle, circle and elliptic can be considered parts of some constellations which the flying saucers can be coming from it.The Priest and the mature Grey has buckets filled mostly with blood of the mayan inside the box.It's a kind of rituals and sacrifices with help of a Aliens.

This Artifact is somehow related with the previous one, we have here the same Grey tall mother or father and we have too a flying saucer with the same form as the previous one in operation of abduction or taking dead alien to his homeland Planet.The taken alien is a kid too.

You feel that this scene is strongly related with the previous one. Triangular, square and circular forms are too presents.The mother or father has a ball with a face this mostly indicate a sacred day of Almanac Calendar with 260 days, based on some Mayan hieroglyphs interpretations. It can be seen too as a sacrifice ritual but for Alien with Mayas mythology culture and if this is true or can be admitted as true in future, this reflects the influence of Mayas on Aliens culture! Finally, the entire ritual can be originated from Grey Aliens then transferred to Mayas or simply it was a sad day for an Alien death that transferred to his tomb in his planet or other place.

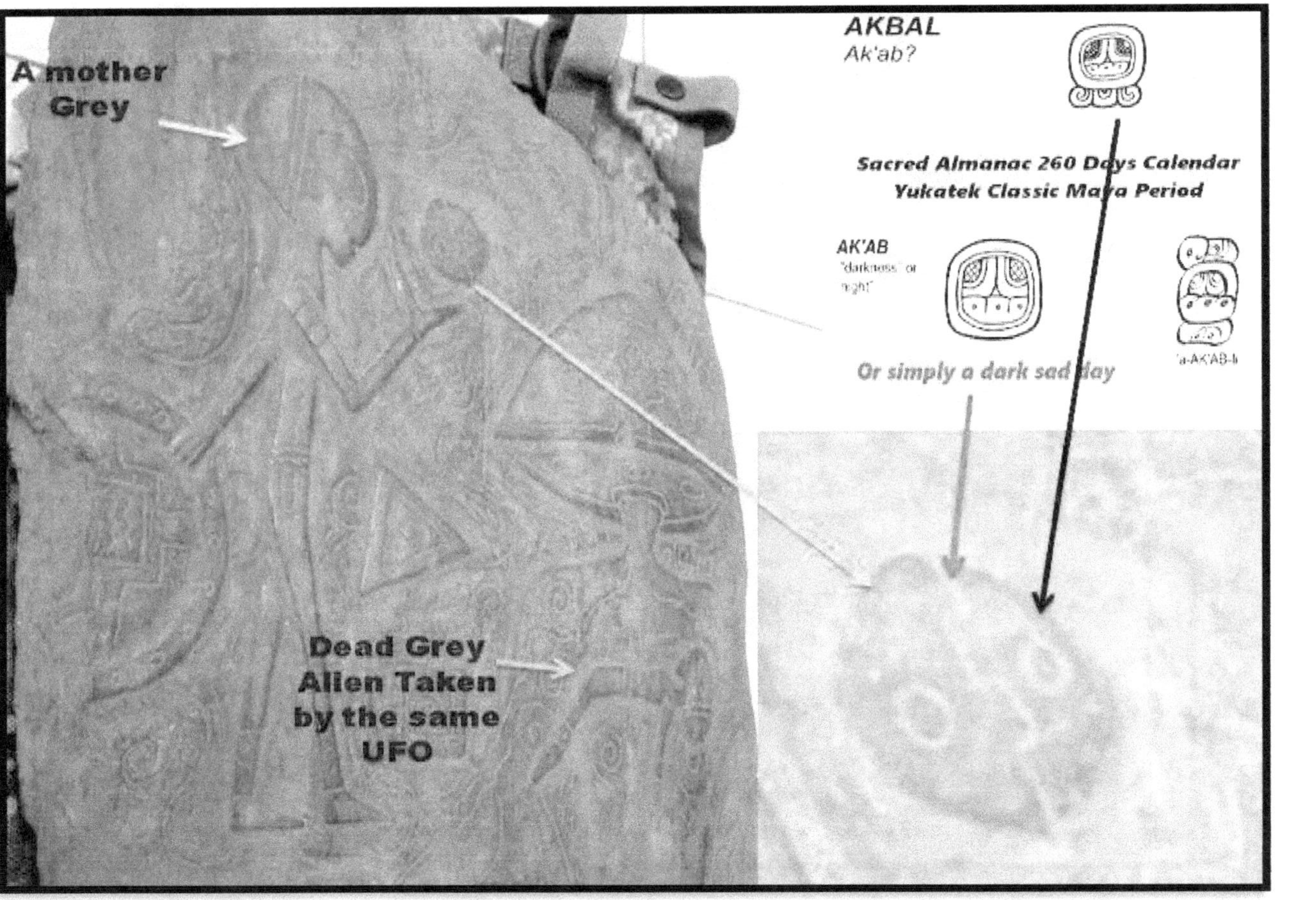

A mother
Grey
AKBAL
Ak'ab?
Sacred Almanac 260 Days Calendar
Yukatek Classic Maya Period
AK'AB
"darkness" or
"night"
a-AK'AB-li
Or simply a dark sad day
Dead Grey
Alien Taken
by the same
UFO

Grey
Alien
holding
Gift from
Priest
Round
flying
saucer
in
action
Grey Alien sitting

Flying Saucer Artifact

A female or male Maya on her or his knee giving a fruit as a gift for a young small Goddess

CHAPTER IX: MAYAN'S PYRAMIDS WITH MICA & MERCURY

There's continuity to unexpected discoveries concerning the ancient Mayans contact with extraterrestrials or seen by Archeologists as ancient gods. The pyramids at Tikal City are another mysterious discussing enigma that adds to the intrigue; reason why, that they are forty kilometers away from a source of water. Generally, the ancient civilizations through history build their cities on or near a large available source of water; as known, Tikal potentially attained the 100,000 citizens, the Mayans would have gathered water from rainfall and stored it in huge reservoirs for use, but logically speaking, why would they leave an easy lifestyle way to build a massive city and do it in a far away inconvenient location?

These pyramids were built throughout the Mayan empire, but 60 of them are found just in Tikal alone. The more well-known Mayan pyramids, including the Temple of Quetzalcoatl, the Temple of the Sun and the Temple of the Moon existing at Teotihuacan City. These pyramids align with the three stars in the belt of the Orion constellation, just like the pyramids at Giza.

A layer of Mica also was found in the construction of the pyramids, the Mica was not available in Central America, it must be found and imported from the region of South America, Brazil of today for an approximate distance of 4000 kilometer away and at that times it were not any modern means of transportation used by Mayans like the wheels. Mica is highly conductive material and is a part of process made of micro-electronics.

The major uses of mica are in sheets and blocks, mica is used as insulating materials in electronic equipment, thermal insulation, gauge glass, dielectrics in capacitors, windows in stove and kerosene heaters, insulation in electric motors and generator armatures, decorative panels in lamps and windows, field coil insulation, and magnet

and commutator core insulation. Specialized applications for sheet mica can be found in aerospace components in air, ground, and sea-launched laser devices, missile systems, radar systems, and medical electronics, making its presence all the more intriguing. The arguing question here is did aliens helped the Mayans to transport this conductive material to the pyramids, and for any specified use?

Another interesting and surprising discovery, a Mexican researcher Sergio Gómez announced that he had discovered "large quantities" of liquid mercury in a chamber below the Pyramid of the Feathered Serpent, the third largest pyramid of Teotihuacan.

While traditional archeology saw this as parts of rituals of ancient Mayans' death mythology especially the mercury is very poisonous and represents a defensive wall to any intruders to the Tombs where it was used, another modern point of view, think that it may have been used as a fuel source for the ancient flying saucers or the so called ancient alien gods' technology.

Some information leaks of our days, confirms that one of energy sources used as fuel for flying saucers is the mercury.So if the Mica material and mercury are used both in the same time in the same pyramid, did this can give a conclusion that this pyramid is a source of charging of flying saucers to make long travels as mica is a conductive mineral and mercury is also a conductor of electricity and did the ancient Mayans go to such great distances to store a lake of toxic mercury under their pyramids just to honor their deads or perform their ceremonies? Or did they using it to empowering an advanced technology?
Three flying saucers, two are rounds and one is cylindrical over a Mayan pyramid, filmed by a tourist in February, 13th 2012 before the end of Mayas Calendar 5126 year.What can be the message that they want to transfer from making this show which is they turn around the cylinder?

Three Flying Saucers over a
Mayan Pyramid dancing!
13/02/20

Zooming!

CHAPTER X; DR GONZALES PRIVATE ALIENS ARTIFACTS

Dr. Gonzalo Franco Martinez is a doctor at the National Autonomous University of Mexico (UNAM), he lives in the city of Teocaltiche, Jalisco, for more than twenty years now.

In his collection of pre-Columbian artifacts there are tens of amazing objects with images of different types of Flying Saucers and Aliens. According to his reports, all these objects were found in one of the archaeological areas of the city of Teocaltiche called "Cerro de los Antiguos".

Teocaltiche is a small town in the state of Jalisco, in west-central Mexico, with more than 25,000 inhabitants. The city was founded in 1530 and is one of the oldest settlements of Hispanic influence at the time of the Spanish conquest. It is officially believed that the first Indian settlements were created in the 12th century. Based on the hill now called "De los Antiguos". The city name Teocaltiche is translated from the Nahuatl language as "a

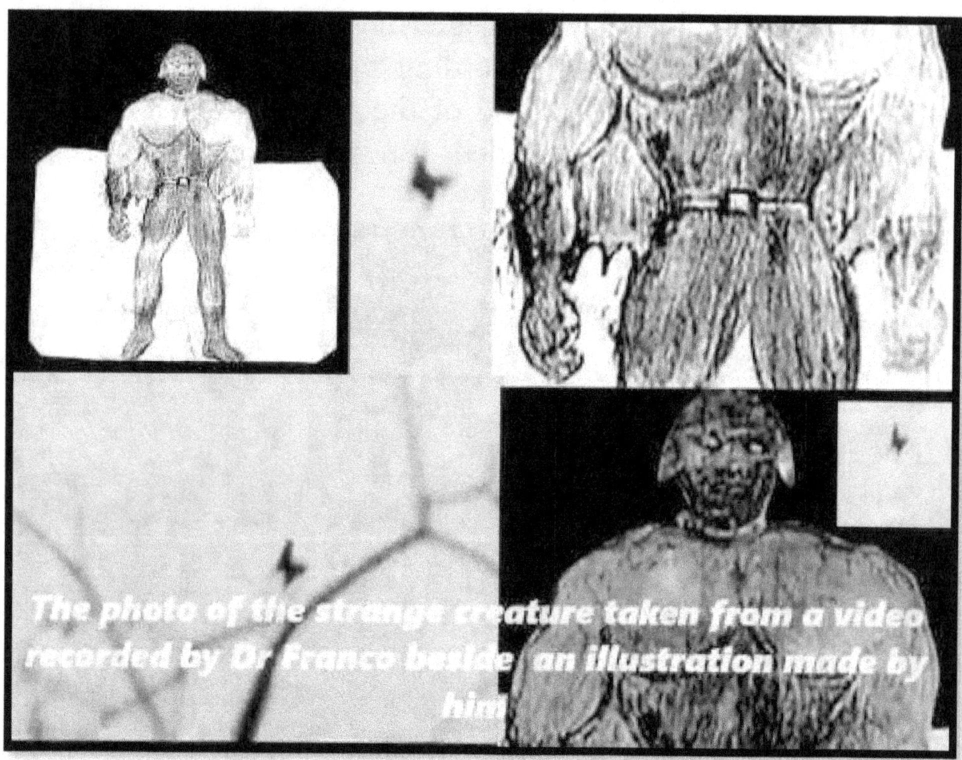

place near the church". Also, there is another version of this name.

Teocaltiche is famous for observing aliens. It is known that in this city everyone has been observed at least once in his life by a Flying Saucer. Dr. Gonzalo Franco Martinez, has been documenting the Unidentified Flying Objects phenomenon for more than twenty years. He made a large number of video recordings of his observations.

In this area of Jalisco, he very often observed flights and landings of different types of Flying Saucers. He believes that in this place there is an energy portal that the aliens use as a space base. One of his registration videos a strange creature with a human and a bat shape appearance in the same time. This is a photos taken from a video and a drawing made by Doctor Franco:

It is important to note that in the doctor's collection presents ceramic artifacts, which allows to conduct their dating by Thermoluminescence method. It is also worth noting that the town of Teocaltiche is relatively close to Ojuelos(approximately 100 kilometer), Jalisco, where, as it's known, famous artifacts with images of Flying Objects and Aliens were found in them. El Toro hills.

The stones of Ojuelos Jalisco in Mexico which show extraterrestrial beings and humans are true, ancient and therefore demonstrate that, in Antiquity, the Nahuatl, the Huachachiles and other wise men of Mesoamerica coexisted with extraterrestrials, the Results and the Investigations were displayed at the 2018 Tepoztlan Unidentified Flying Objects Congress.

One of teams who visited Dr Franco was a Russian team who have carefully examined the collection. It's known that local medieval ceramics are widely existing and spread, but it is interesting to note that most ceramic objects, representing a variety of extraterrestrial planes, were found in the form of "fragments" in Cerro de los Antiguos hill, in the center of the city.

There is an assumption that the hill itself is invaded by the ancient pyramid. Indeed, for this; there are some fields contains a hill with 25 to 30 meters high, pyramidal in shape and there are often flat carved stones that which could belong to the pyramid. The Russian team was convinced of this during a visit to the hill.

The entire top of the hill is dotted with holes and everywhere you can find fragments of old pottery.They sent some artifacts to the well known and credible German Laboratory Kotalla for the Thermoluminescence tests, but What is Thermoluminescence?When a small sample of ancient pottery is heated it glows with a faint blue light, known as thermoluminescence or TL. During its lifetime every pottery absorbs the radiation from its environment something which creates thermoluminescence on it.

The older the pottery, the more radiation it has absorbed and the brighter the pottery sample glows. By measuring the TL, it can be calculated how much radiation has been absorbed and use this information to calculate the approximate age of the pottery.This labo, Kotalla, is working with high ranking universities, institutions, Museums, etc, so it has great credibility as it has great responsibility towards his clients.

For eample, the Artifact below showing a Flying Saucer has an edge between 892.5 and 1207.5 years old.

Flying Saucer Artifact with an age between 892.5 and 1207.5 years old with Thermoluminescence Test

Another kind of tests were applied on other Artifacts which is the Radiocarbon Analytical test from the laboratory AMS of Arizona University . Radiocarbon dating is a method that provides objective age estimates for carbon-based materials that originated from living organisms. An age could be estimated by measuring the amount of carbon-14 present in the sample and comparing this against an internationally used reference standard.

A pipe showing Aliens Greys, planet, Flying Saucer, etc.

THE UNIVERSITY OF ARIZONA

UNIVERSITY OF ARIZONA
AMS LABORATORY

10676) – Radiocarbon Analytical Report

Data Report (1 of 1)

User Information	Laboratory Information
Submitter:	AA-number: AA110676
User ID: organic adhesive	Laboratory number: X32399
Expected age: ~5000 years old or older	Sample type: adhesive
Sample origin: Ojuelos, Jalisco Mexico	Pretreatment: ABA
	Carbon yield: 37.8%
	Carbon mass: 1.41 mg

Results

$\delta^{13}C$ (± 0.1‰, 1σ):	-28.5 ‰
Fraction of modern carbon (±1σ):	0.3000 +- 0.0011
Uncalibrated ^{14}C Age (±1σ):	9671 +- 30 14C years BP
Calibration Program / Dataset:	OxCal 4.2 / IntCal13 atmospheric
Calendar Age Range (68%):	9232 calBC to 9146 calBC
Calendar Age Range (95%):	9249 calBC to 8922 calBC

A cmparison between Doctor Gonzalo collection of Flying Saucers and the sightings made by peoples through modern history give these:
Typical forms between twentieth century sightings and the supposed Mayan's Artifacts:

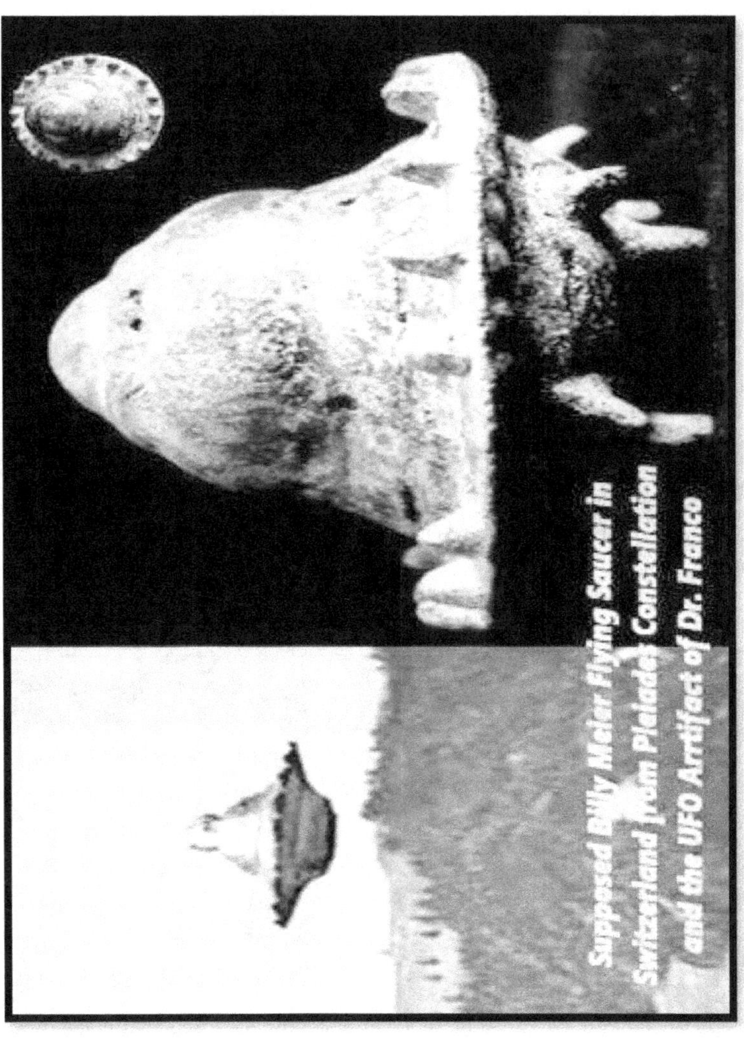

Same Forms and
same Shapes in
comparison with
Dr Franc
Artifactz
Same Form between the Mayan's Flying
Saucers and the ones of our days

Similarity to the modern
Flying Saucers

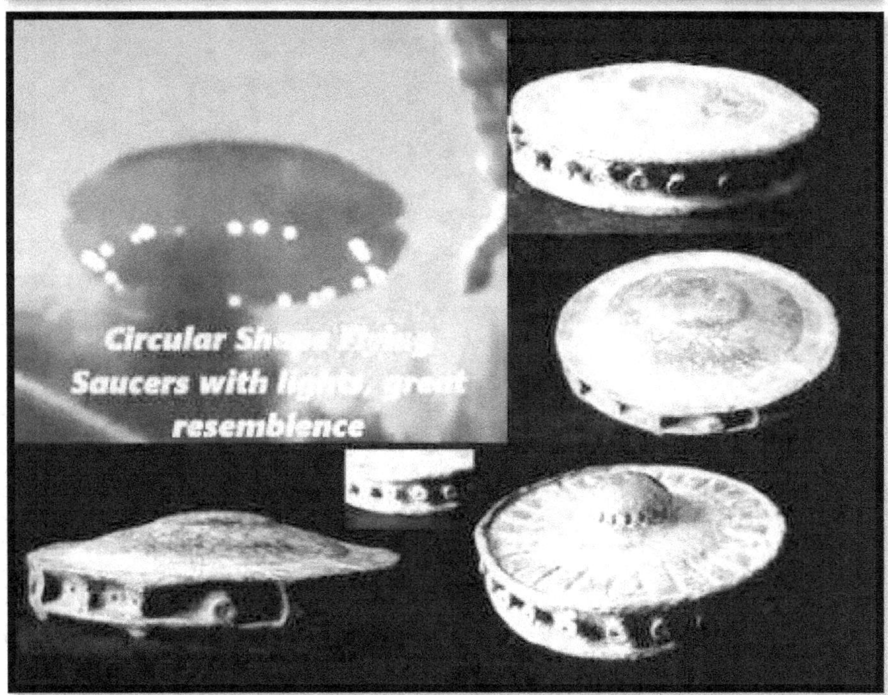
Circular Shape Flying
Saucers with lights, great
resemblence

Mayan Artifact
and a Flying Saucer taken over a US Park
Strange Similarity

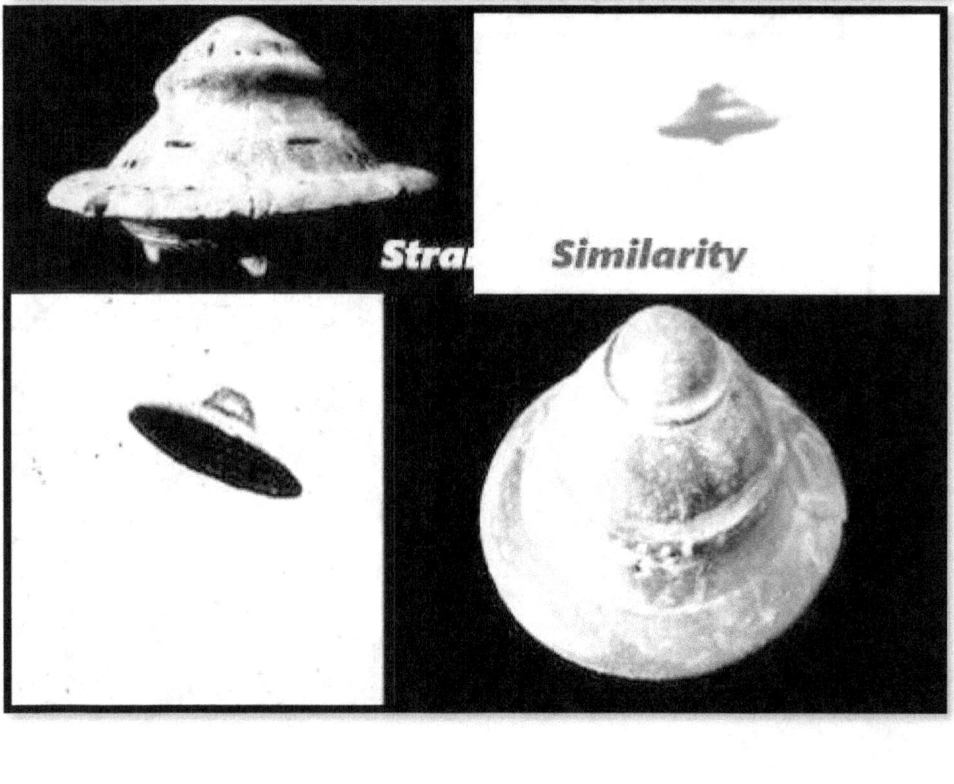

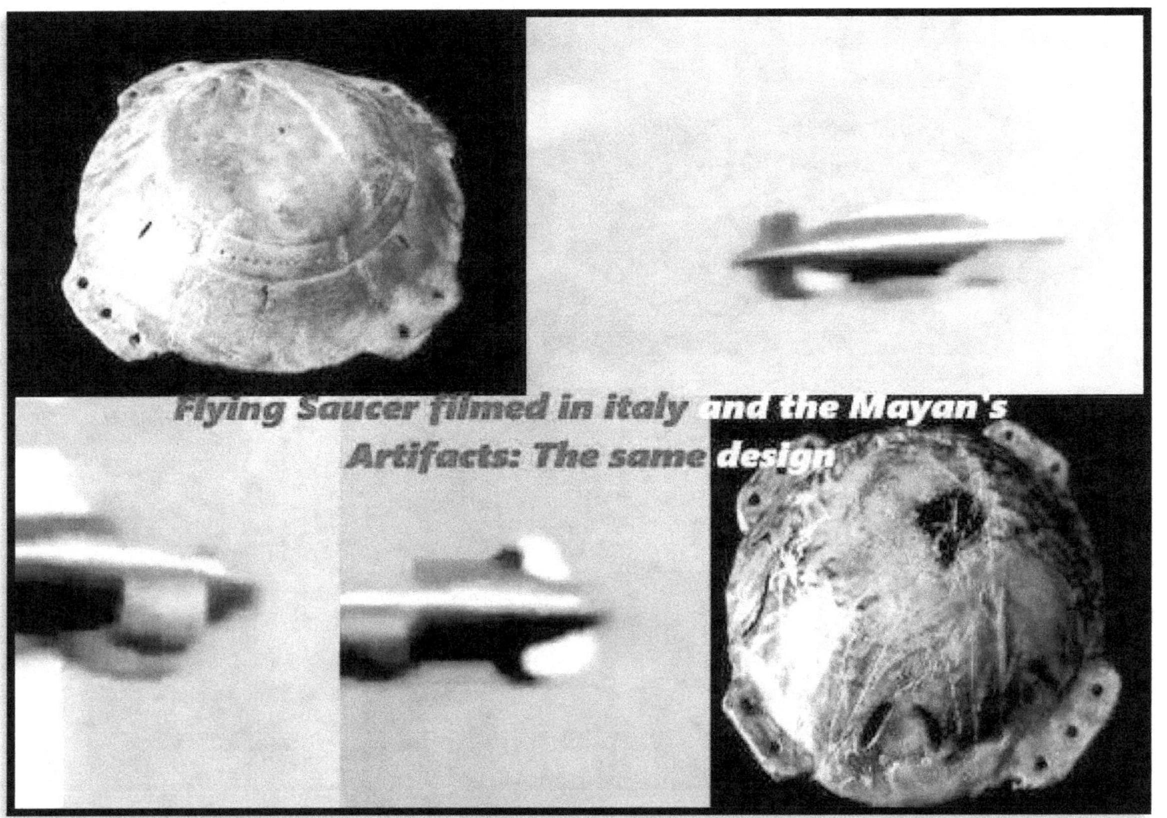

Flying Saucer filmed in italy and the Mayan's
Artifacts: The same design

Another strange similarity is between so many cigar shape Flying Saucers with lights windows throughout seen in the last and present centuries:

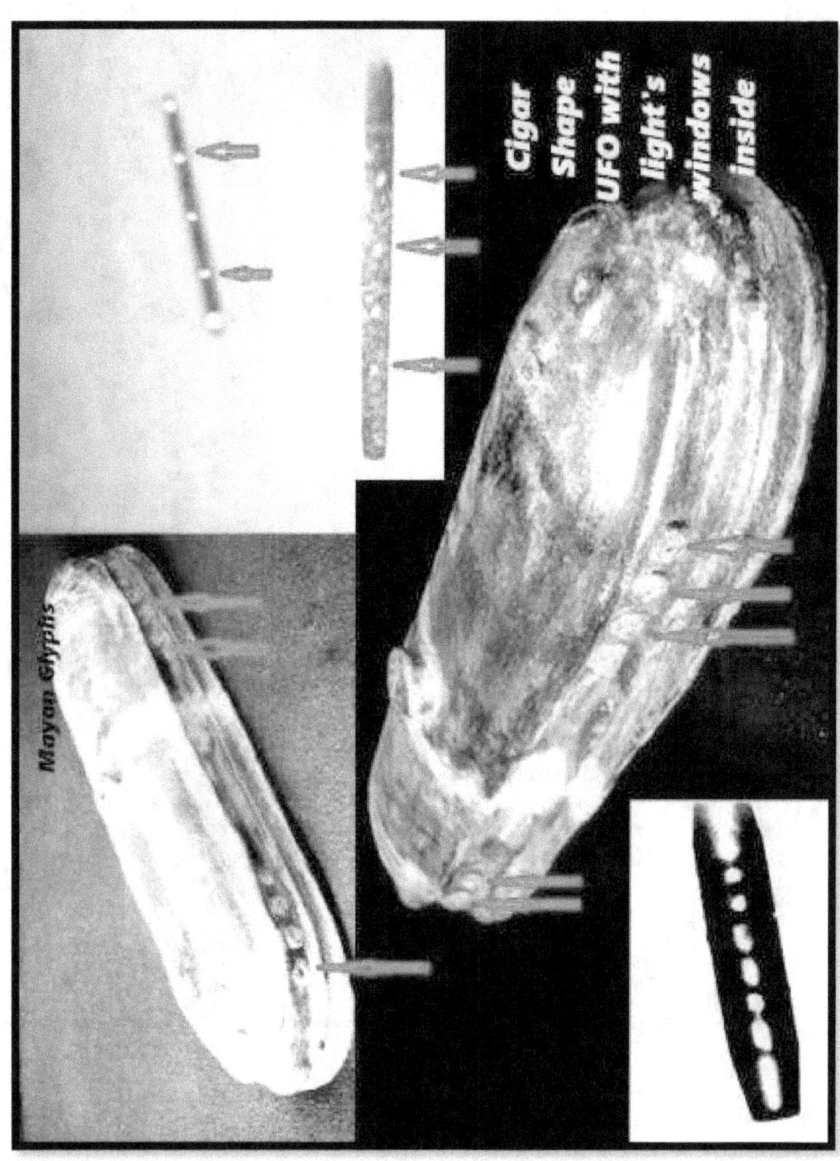

The two axes Spaceships or Flying Saucers are too presented in Doctor Franco Collection.Someones of our modern time are with curved angles. One of these ufos was photographed in 2008 by a Turkish citizen in Turkey with his amateur telescope, it was conducted by two Greys.

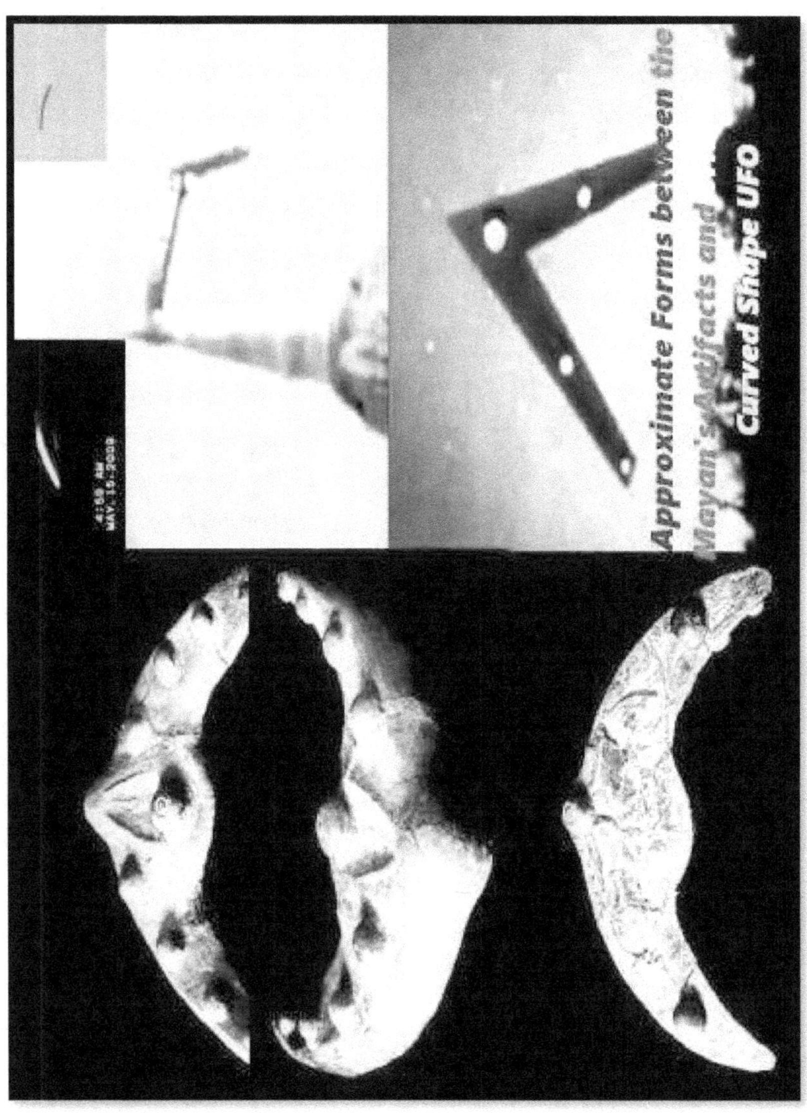

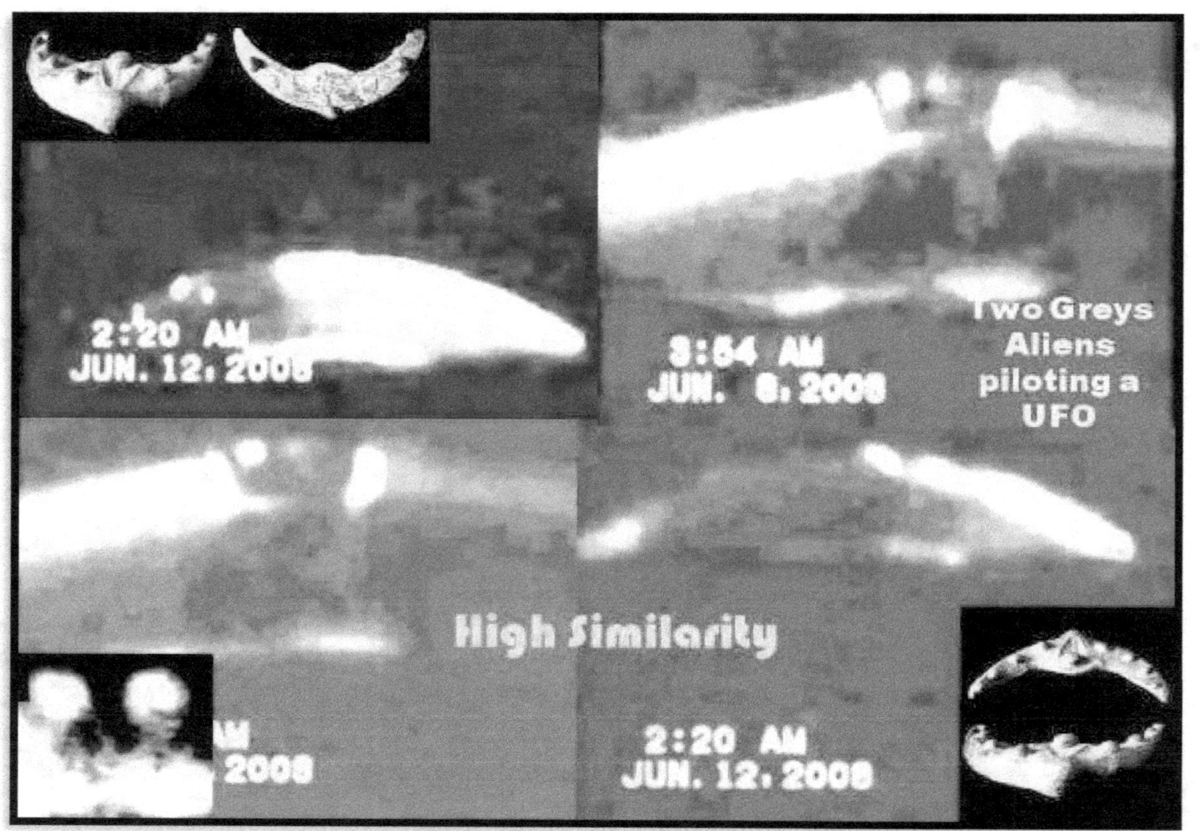
2:20 AM
JUN. 12. 2008
3:54 AM
JUN. 8. 2008
Two Greys
Aliens
piloting a
UFO
High Similarity
2:20 AM
JUN. 12. 2008

A water container or it may be considered as a dinner plate presenting Flying Saucer the same as shown for the twentieth century secret UFO and an Alien with horns, the same discovered in Doctor Gonzalo Franco Collection:

The picture below shows a Statue with a Grey Alien with oval eyes with great skull. Drawings of a Mayan in front as he cry towards this Alien as a flow of air emerging from his mouth reflecting the amount of crying that he put.In the back of the Statue a Mayan with a form of a Bird sitting.In his back three stars resembling the ones of the Orion constellation, the Belt. Over him, a Flying Saucer traveling towards a planet.

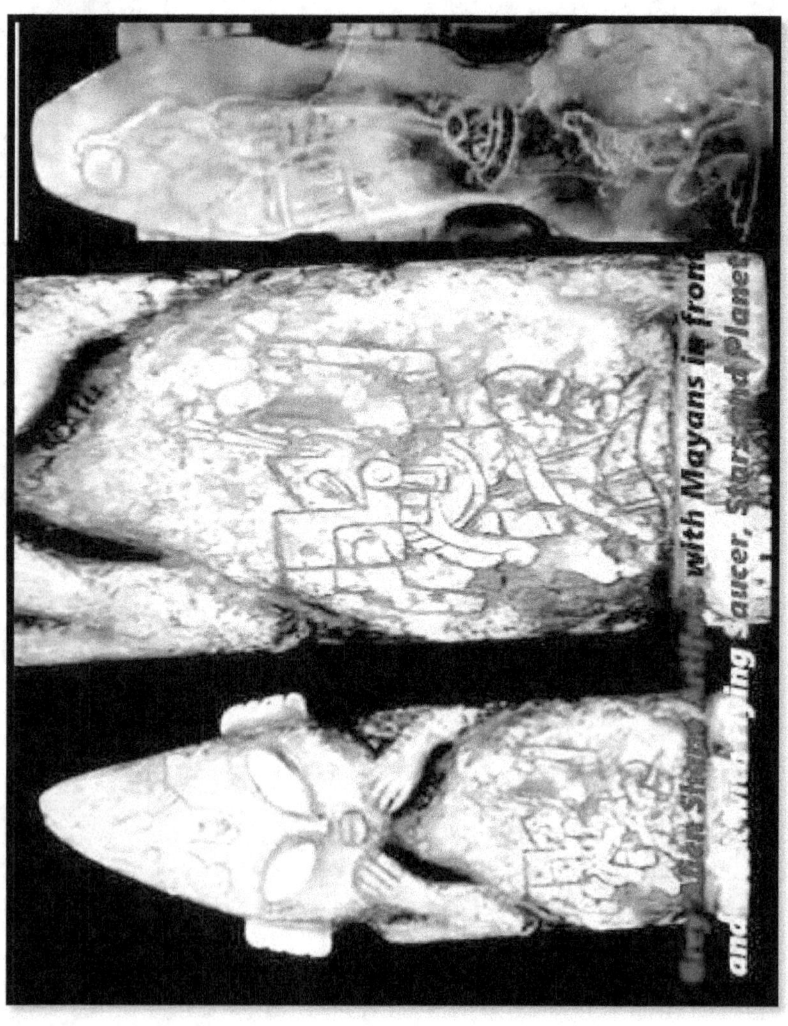

A Stone Artifact showing a Grey Alien descending from a Flying Saucer and give a Mayan something. The Mayan has a pipe in his mouth. It can be a gift or a medicine to this Mayan for therapeutic use because he is in a kind of traveling machine interpreted as the travel to death by some Archeologists.In parallel there are feet prints, small flying Saucers and a Planet

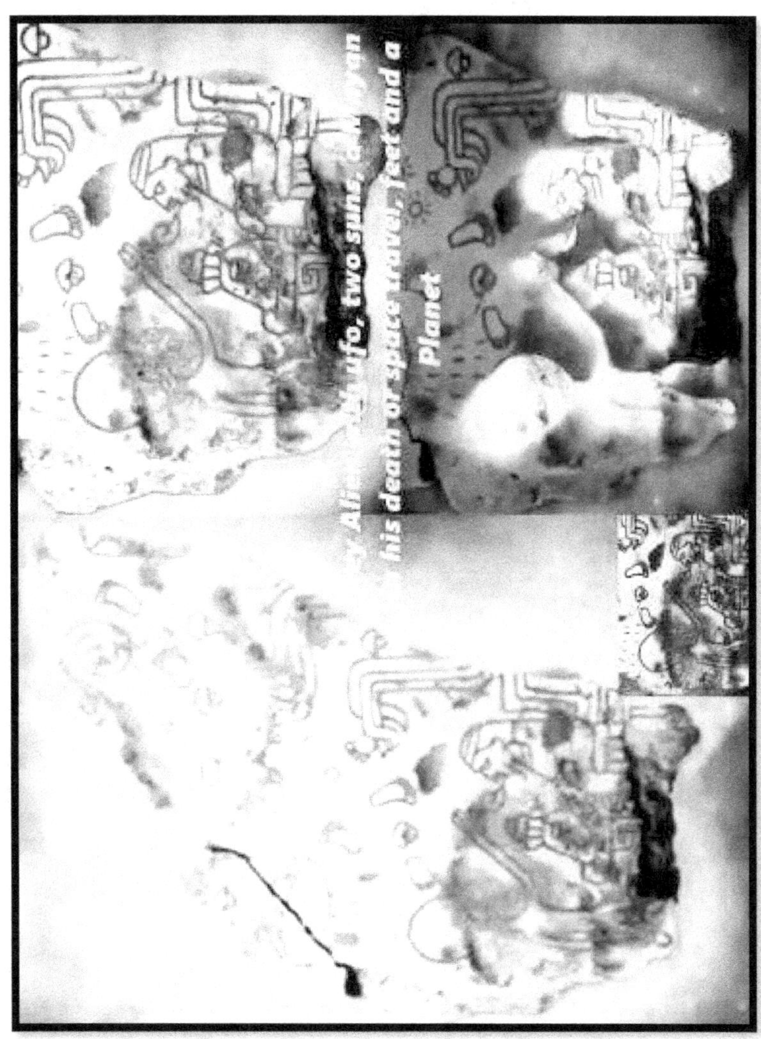

I found the feet prints in another small Artifact of a Grey Alien.The feet are with five toes not three as other artifacts.So the foot prints can be considered as a proof of existence or visitation of Aliens to this land, the Mayan land.

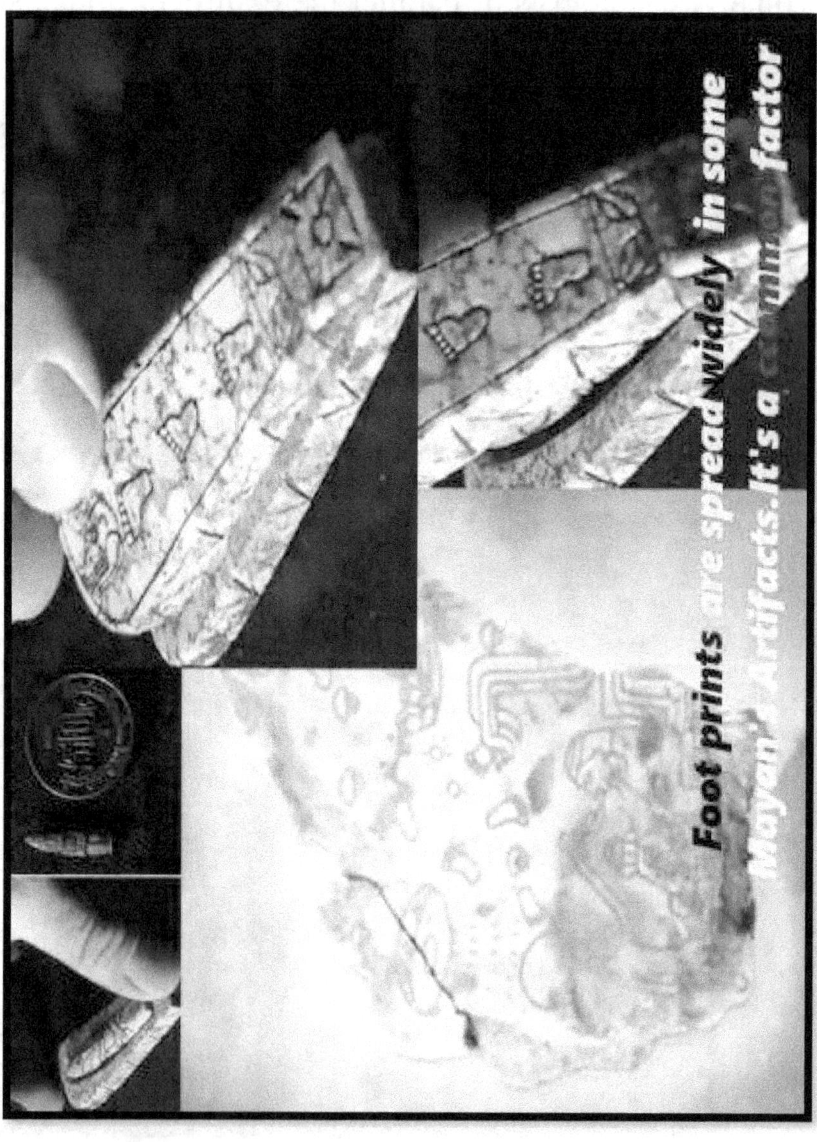

The Collection of Doctor Franco presents so many statues.When we take some of them and compare with some aliens or extraterrestrials uncovered now for decades by the truth seekers and the ex-secret programs Agents whose gave great push to this subject.The image below give an approximate similarity between a Statue Artifact and an orange color strange Alien not known as the Grey Aliens ones.

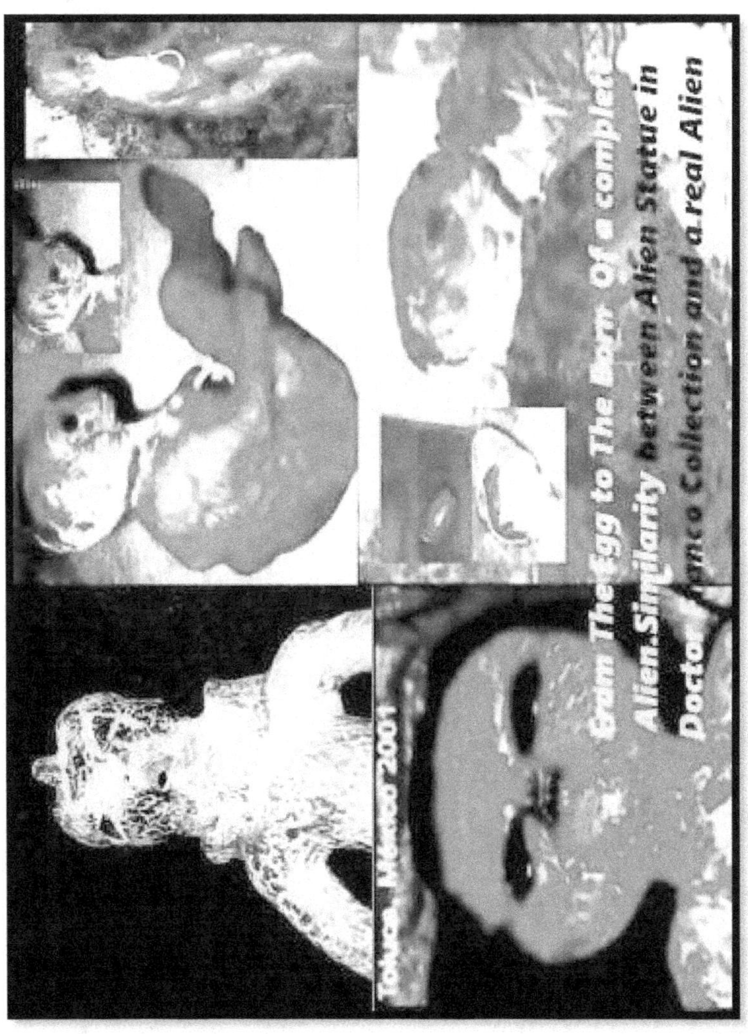

A crying face Artifact with Great eyes.The face has a crown surrounding its head.Over the crown, mostly a dead king will be taken by a Flying Saucer.The screaming face maybe reflects this sad situation:

In the both sides in the image below the Flying Saucers are existing but in one side a Mayan with wings give a gift to an alien emerging from this UFO.The other side a Mayan came with a Flying Saucer over him an people receiving him with happiness, has too wings and give the feeling that they want to fly.

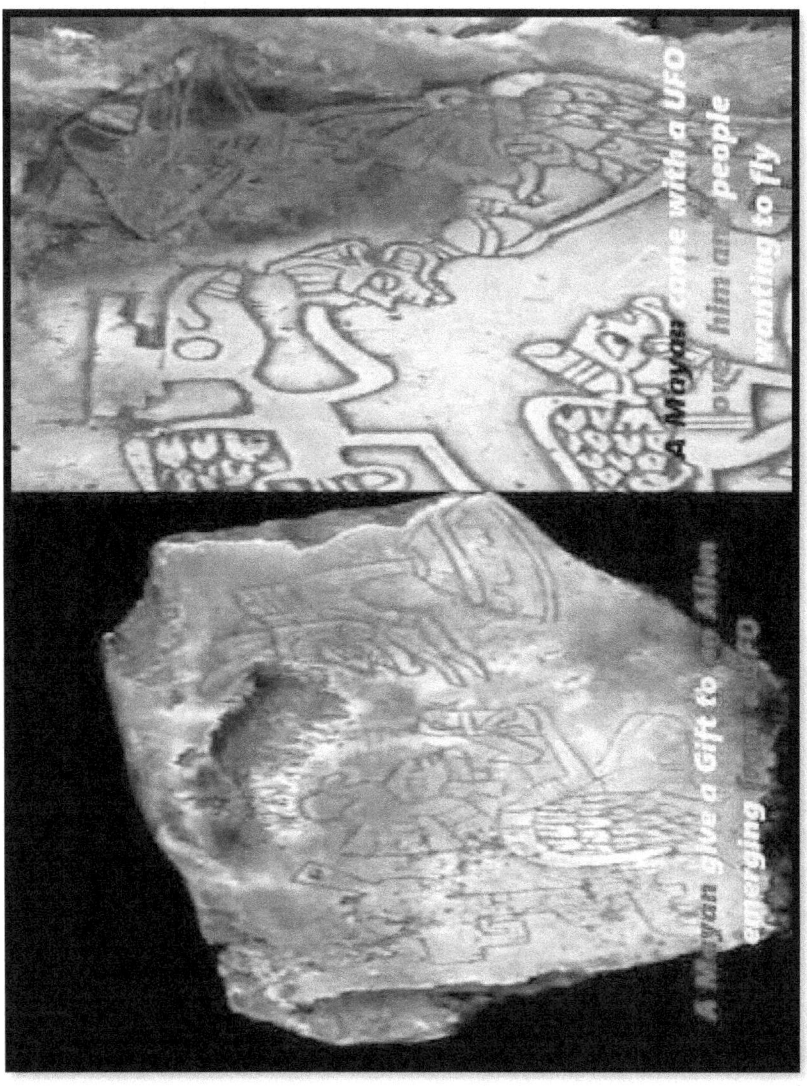

A Grey Alien with wings holding a ball it could be the ball game from Mayans used in Ballcourts and took it as a gift.The right picture show a Mesoamerican which could be Mayan or not holding a Grey Alien dead or alive or it could be a Grey Alien as his eyes are large.

In the image below, we have a Spaceship in the form or design of a Whale.We see windows in both sides and a front piloting windows.This supposed Spaceship has a propulsion system.

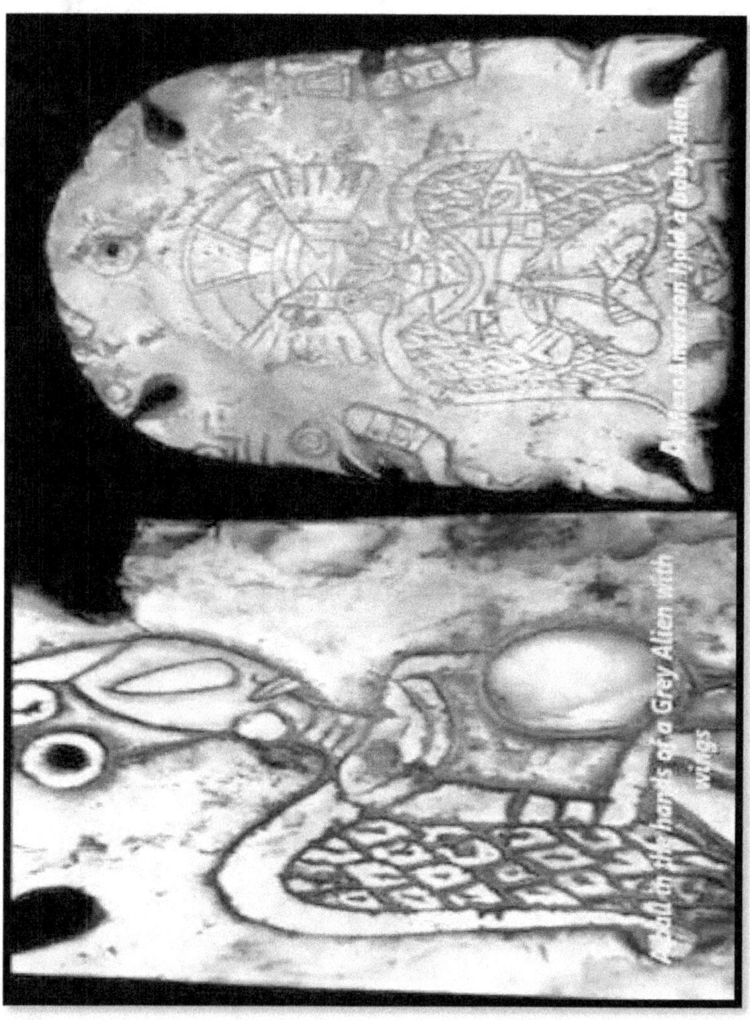

A Fish Form Spaceship also found in Doctor Franco
Collection:

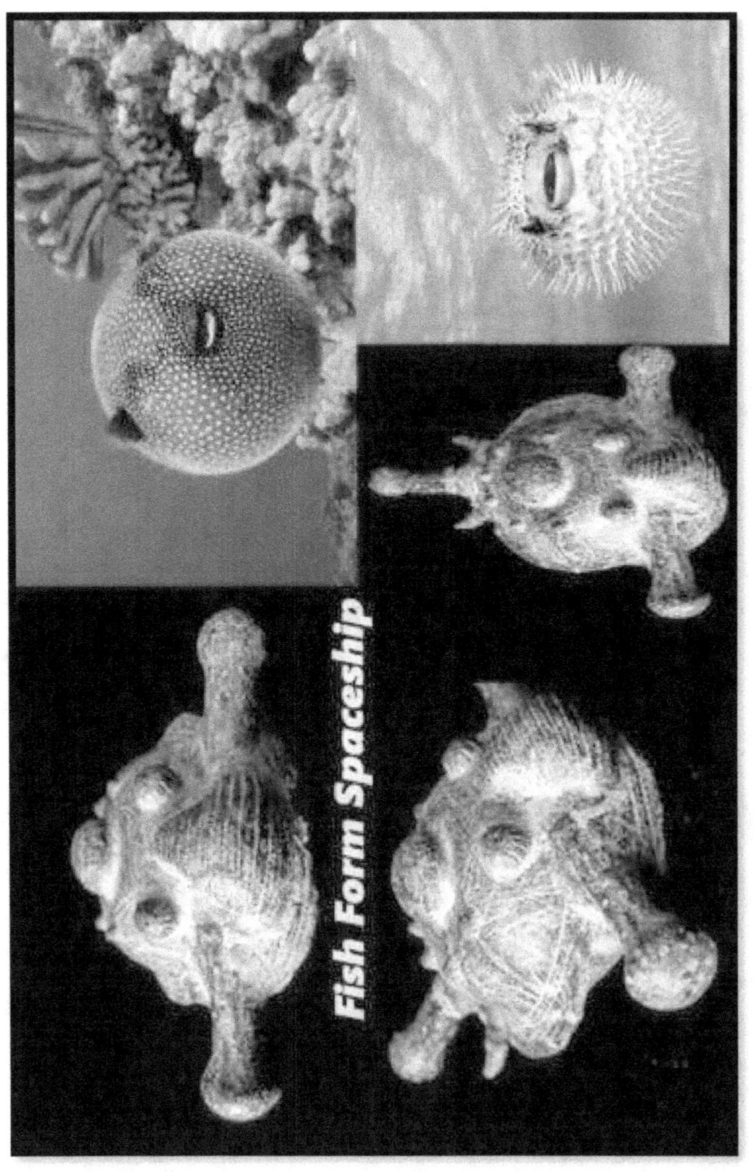

The Grey Alien Form is in abundance existence in the majority of Doctor Franco Collection:

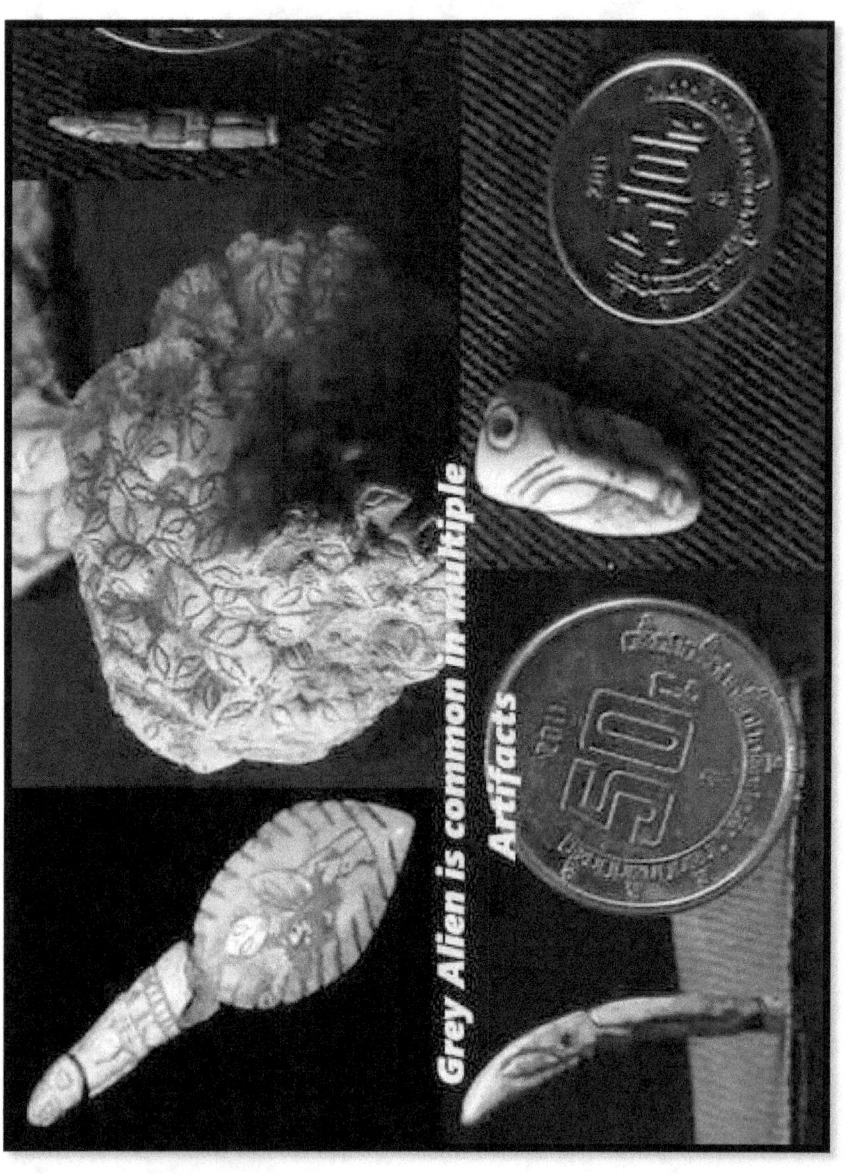

Artistic Statue, very similar to the
Statue of Doctor Franco
Collection.Coincidence?

CHAPTER XI : KING PAKAL SPATIAL TOMB

SakK'inich Janaab' Pacal known as " Pakal The Great", was born on March 23, the year 603 , the son of Lord K'an Mo'Hix and Lady Sak K'uk', the reigning Queen of Palenque. Known as the lady Beasty for Archeologists, the Queen Mother was one of few women of Pre-Colombian dynasties that ruled. King Pakal, whose name means "Sun Shield" in the Mayan mythology, was crowned by his mother on July 29th, the year 615, shortly after his 12th birthday.

He was known for his constructions and extensions of buildings and monuments like the Palace of Palenque, one

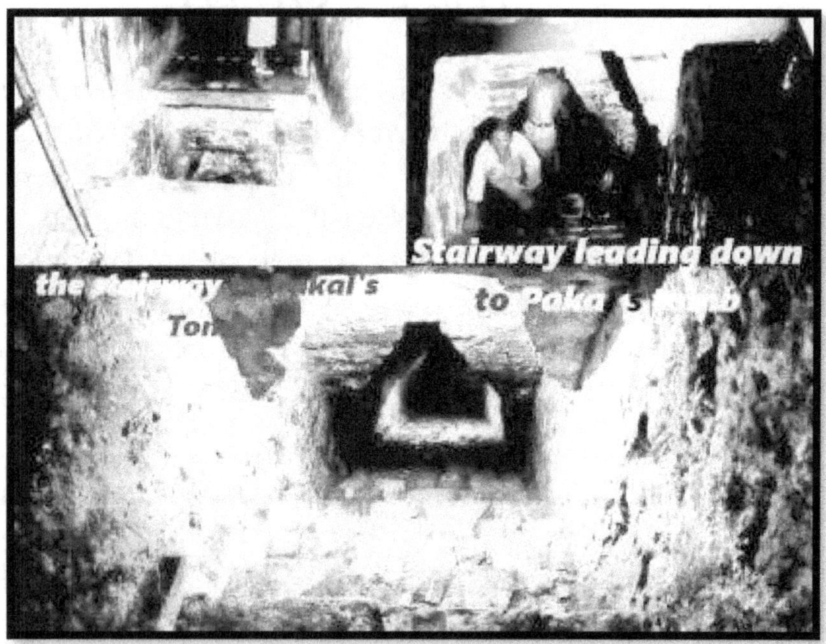

of his widely famous monuments, the temple of inscriptions beside the Palenque Palace. He reigned for 68 years which is considered as one of longest governing period in the Mesoamerican cultures. He died at the age of 80 in August 683.

Pakal was buried in a colossal sarcophagus in the Temple of inscriptions which means also in Maya mythology "The House of the Nine Sharpened Spears». Despite Palenque had been uncovered for long years and examined by archaeologists, the tomb of Pakal was not discovered until the year 1948 when the Mexican Archeologist Alberto Ruz Lhuillier did it.

It
took
four

years to clear the rubble from the stairway leading down to Pakal's tomb, but it was finally uncovered in 1952. His skeletal remains were in his coffin, wearing a jade mask and bead necklaces, surrounded by sculptures and stucco. Traces of pigment show it was painted with colors, which is a common factor of Maya sculptures.

CHAPTER XII : THE GREAT SARCOPHAGUS LID OF PAKAL

Pakal's sarcophagus lid is made of a single stone. It is a rectangular form, measuring between 245 and 290 millimeters (approximate: 9-11.5 inches) thick in different places. It is between 2.2 meters wide by 3.6 meters long (about 7 feet by 12 feet). This huge stone weighs seven tons.

There are Mayan's Hieroglyphs on the top and sides. It would never have fit down the stairways from the top of the Temple of the Inscriptions to the Tomb. Logically thinking, Pakal's tomb was put first and then the temple was built around and over it. When Ruz Lhuillier discovered the tomb, he and his team carefully lifted it with four jacks.

The reborn conception or the reincarnation is a basic and principle interpretation for a majority of ancient civilizations and the Maya is one of them. After his death and the death of all his parents, they travel to the eternal life in other forms that's why so many Archeologists did the same for the Glyphs on the Pakal's Tomb and especially the lid of Sarcophagus.

The Glyphs of the sarcophagus lid talk about events from the Pakal's life and some of his royal ancestors. there are eight fascinating carvings of Pakal's ancestors being reborn as trees.The southern edge records the date of his birth and the date of his death. The other borders mention several other lords of Palenque and the dates of their deaths. The northern edge shows Pakal's parents, along with the dates of their deaths.

Those Glyphs are considered as images and those images are separated as:

Two images of Pakal's father, K'an Mo' Hix, being reborn as a nance tree, two images of Pakal's mother, Sak K'uk', being reborn as a cacao tree,Pakal's great-grandmother, Yohl Ik'nal, is shown twice, reborn as a zapote tree and an avocado tree, Janahb' Pakal I, Pakal's grandfather, reborn as a guava tree, Kan B'ahlam I (ruler of Palenque 572-583), reborn as a zapote tree, Kan Joy Chitam I (ruler of Palenque ca. 529-565 A.D.), reborn as an avocado tree, Ahkal Mo' Nahb' I (ruler of Palenque ca. 501-524 A.D.), reborn as a guava tree.

CHAPTER XIII :THE TOP OF THE SARCOPHAGUS LID

The top of the Sarcophagus lid is the most discussing subject about the entire Tomb's Discovery.One of interpretations is Pakal being reborn. Pakal is on his back, wearing his jewels, headdress, and skirt. Pakal is shown in the center of the cosmos, being reborn into eternal life. He has became one with the god Unen-K'awill, who was associated with maize, fertility, and abundance.

He is emerging from a maize seed held by the so-called Earth Monster, whose enormous teeth are clearly shown. Pakal is emerging along with the cosmic tree, visible behind him. The tree will carry him to the sky, where the god Itzamnaaj, the Sky Dragon, is awaiting him in the form of a bird and two serpent heads on either side.

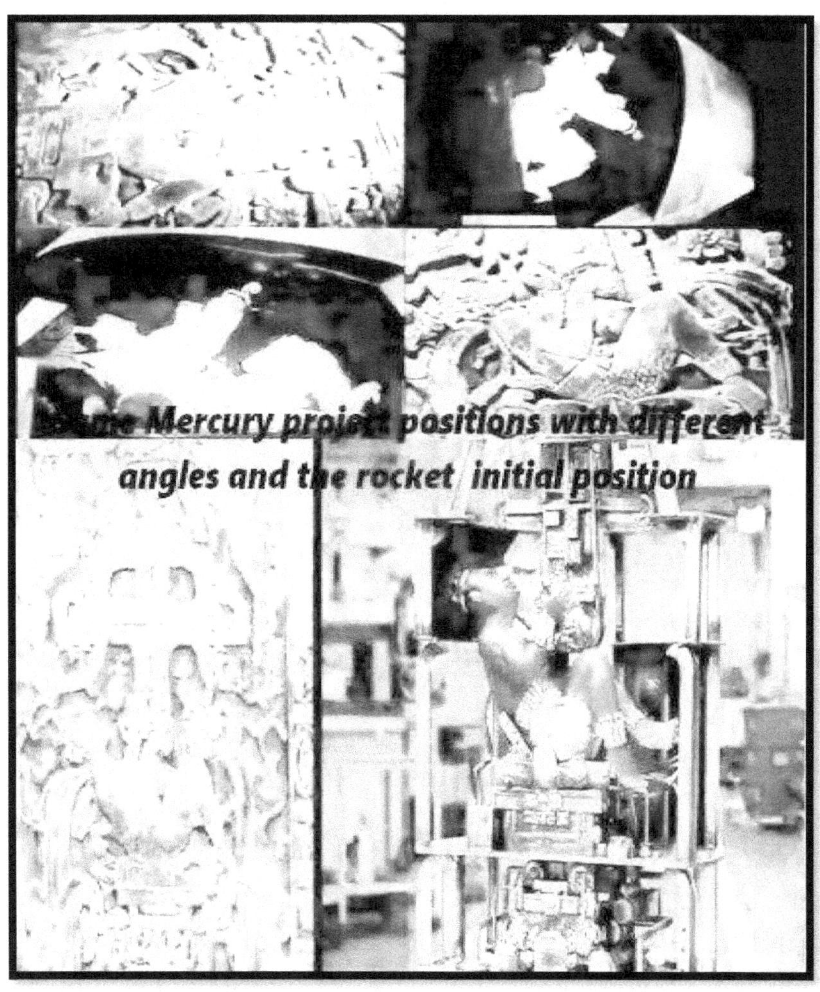

The first who talked about the Ancient Astronauts or Aliens theory is Erich Von Daniken in his book Chariots Of The Gods. He compared between the Pakal lid with the Astronauts position in Mercury space mission in the sixties of the twentieth century. He interpreted this image as a flying rocket. After that many are involved in this same idea.

Te most important and famous symbol is the world tree, based on the ancient astronauts theory this tree is a rocket. The world tree was a symbol key in Maya mythology. Its branches extended into the heavens while its roots are in the underworld. So it's a symbol of bridge between the underworld, the earth and the heavens.

In the same art lid, there's a double headed vision serpent surrounding the world tree.This vision of serpent is thought to leave the center of the world, it's placed between the underworld and heavens.

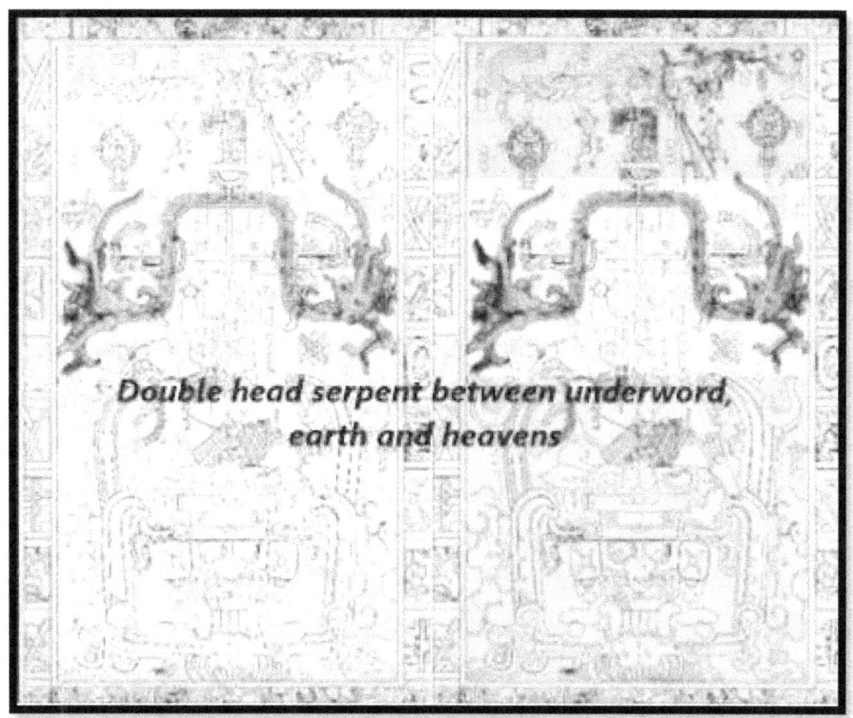

In the tree's top branches, there's a celestial coaxial bird

which is commonly seeing in other Mayan world tree depictions. The celestial bird represents the heavens and that's why it's always put on the top of the world tree.

At the base of the lid, the ancient astronaut theorists interpret it as a fire for the propulsion of the rocket take off, others said that it is just another roots of the world tree extending to the underworld.

In the same lid, we find the sun monster under Pakal directly, in Maya mytnology, every sunset is a travel to the underworld where it dies like everything else.The Sun Monster is composed by a skull and a living flesh upper him presented in hands, it's a common thing between Mayan arts showing the transitions.

The smoke interpreted by the ancient alien's theorists, is considered by this theory as a tool of transition from a state to another or place to another one and it can be changed from a Glyph to another sometimes it's a serpent mouth or umbilical cord, etc.

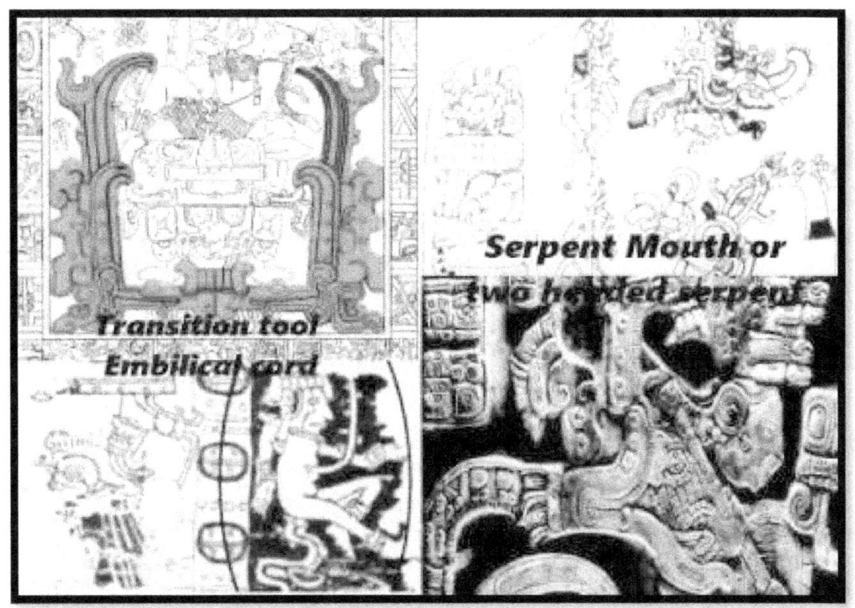

For king Pakal itself, the ancient astronaut theory supporters said that he manipulate the rocket via some keys with his hands left and right, but, the opposes said that his right hand don't touch anything and the left one is manipulating something but that's not means that he manipulated all the controls of the world tree! The hand of king Pakal is a common in Maya's Arts!

For the foot position of king Pakal, it's considered a kind of a pedal to help him to piloting the rocket, the other side opinion said that it's a Mayan art symbol existing elsewhere without a rocket.A tube of breathing in his nose considered as just a nose piercing by a bone.

In conclusion, this point of view see that the complete lid is a reflection of the pass of king Pakal from the underworld to heaven in his death travel and all the story is an art piece.

Another interpretation came from two authors: Adrian G.Gilbert and Maurice M. Cotterell in their book, The Mayan Prophecies. They found what they believe is a hidden code, they unlocked 128 different images or codes from this lid.

A secret instructions are existing in the borders of this lid. These instructions when the sarcophagus lid is rotated with different angles, it gives secret codes in the form of images of Aliens or developed technologies and more Mayan's mythology new symbols never revealed before.

One of pictures or codes showing king Pakal with a sign of a bat in his mouth. The bat is a sign of death in Maya mythology, so is a symbol of taking the king Pakal's breath.The high hair style found in the statue head of king Pakal in his tomb becomes a baby Quetzal bird carrying a chain which is the highest god in Mayan culture. So Pakal became a bird after his death reincarnation meaning.

The authors saw that the lid is a panoramic sequences of life's images in universe. The codes are in parts a Mayan mythology description but in other parts astronomical knowledge.

These codes are also depicted from other Mayan artifacts like the king Pakal's jade mask found in his tomb.
Several interpretations are still emerging, but if we consider that it's at the end a Mayan mythology describing the process of transfer from the actual to the afterlife, the complete table reflects an attachment of Mayans to the sky or heavens culture and what can be seen as technical side of the space voyaging.

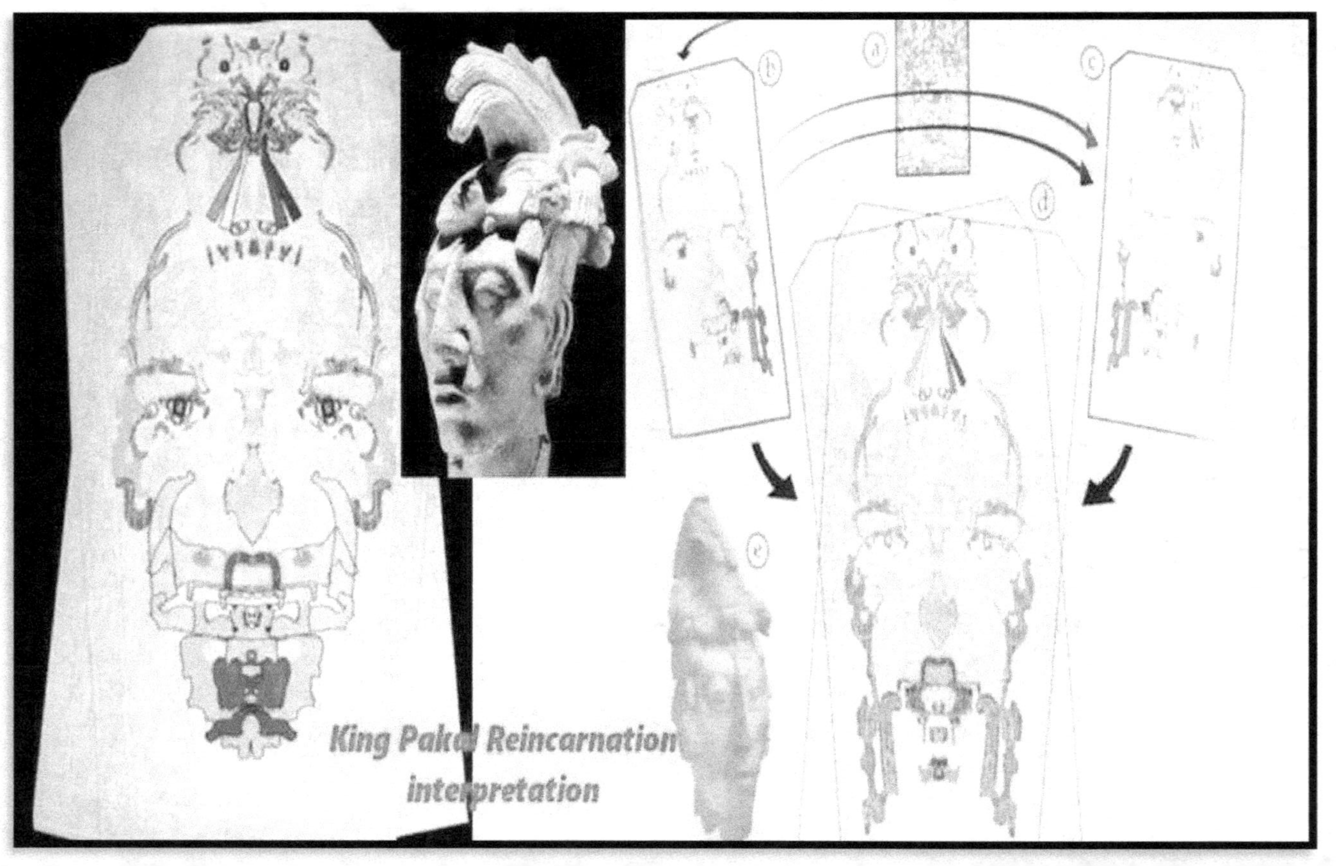

King Pakal Reincarnation
interpretation

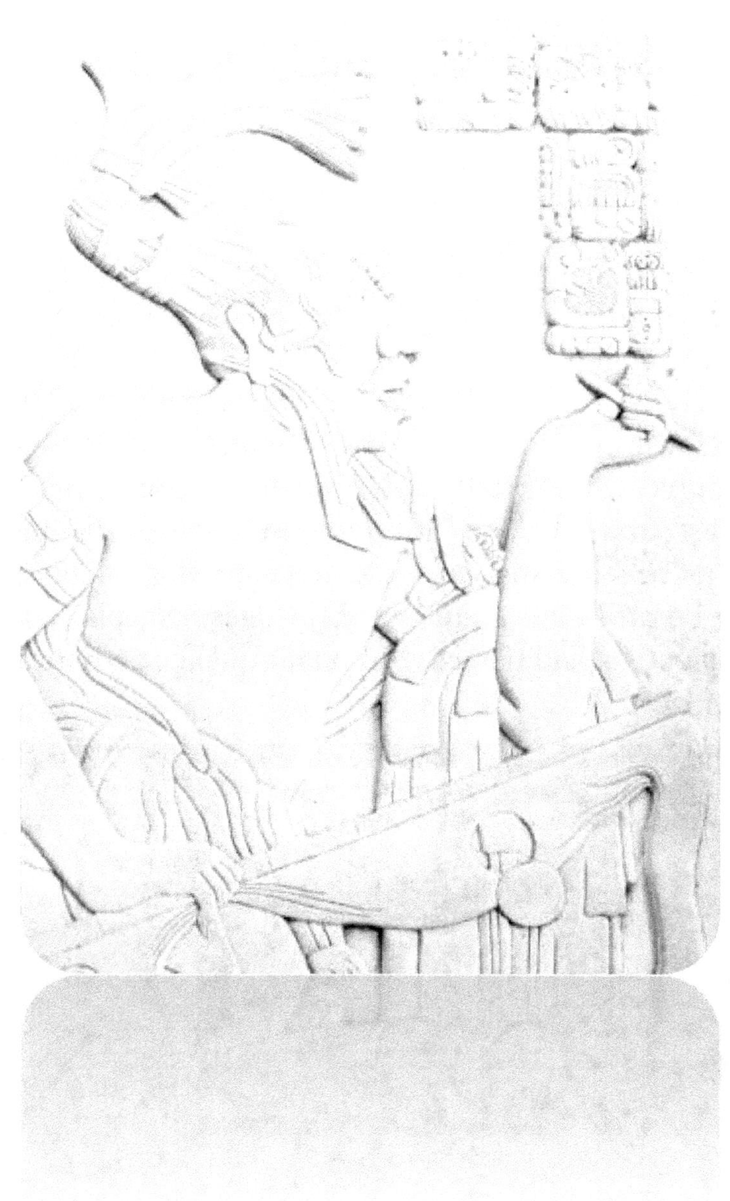

CHAPTER XIV : MOON SPACESHIP AND MAYAS

Apollo 15-Apollo 20

A huge structure was discovered in our moon in one of NASA missions in the twenty century, these missions were the Apollo program, Apollo 15 to 20, with an appearance of a concrete structure, it can be considered as a great Flying Saucer for great space travelling distances in our known universe or a tall and wide tower for a race of extraterrestrials. It can be put down in a war or due to its old age or because moonquakes, not necessary naturals , they can be artificial, a human made ones, especially that NASA scientists had detected so many quakes activities in the moon.

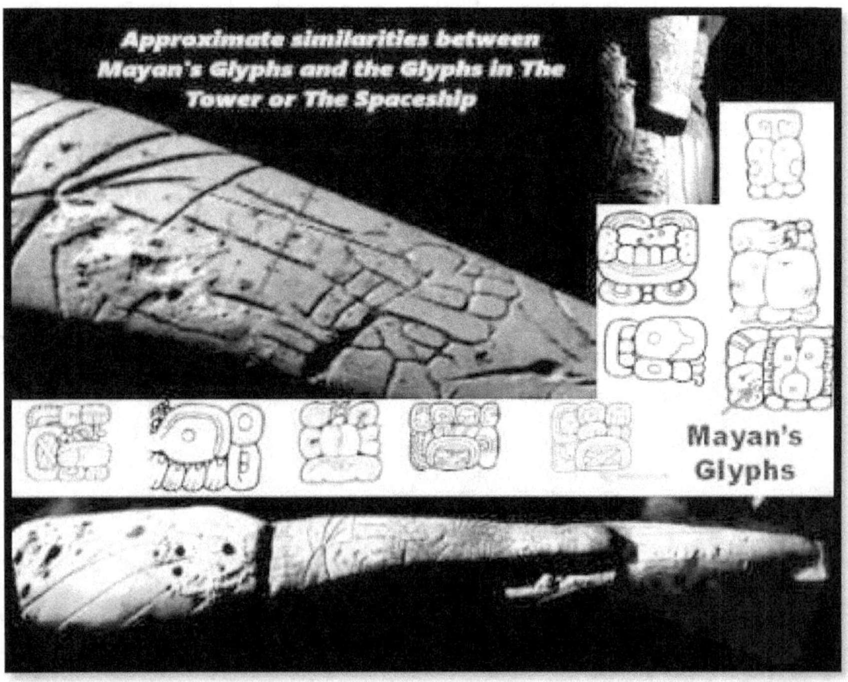

The length of the tower is 3370 meters and the width is 500 meters, it's so enormous in comparison with modern human made structures like Eiffel tower, great seas ships and others

The tower shows several draws similar to the Mayan's Glyphs presented in codices of Madrid, Dresden, Paris and Grolier. Does this reflect a potential relation between Mayas and this Tower or Spaceship? Especially as it's known that the Maya's civilization has a lack of records for more than one hundred years and a strange and mysterious mass disappearing?

Another discovery that I made in this subject is the existence of a human face and head who can be a Mayan, an Indian or a Mesoamerican over this same building or Spaceship!

The supposed Apollo 20 mission was a top secret program that's coming after the fail of Apollo 16 to 19.It's a person called Mr William Rutledge who was an astronaut in the crew during the seventies of the twenty century employed by the USAF in collaboration with NASA during a secret space mission. William Rutledge is a man with 76 years old at the time of his speech about this story, who lives in Rwanda. He is an American citizen now, civilian, born in

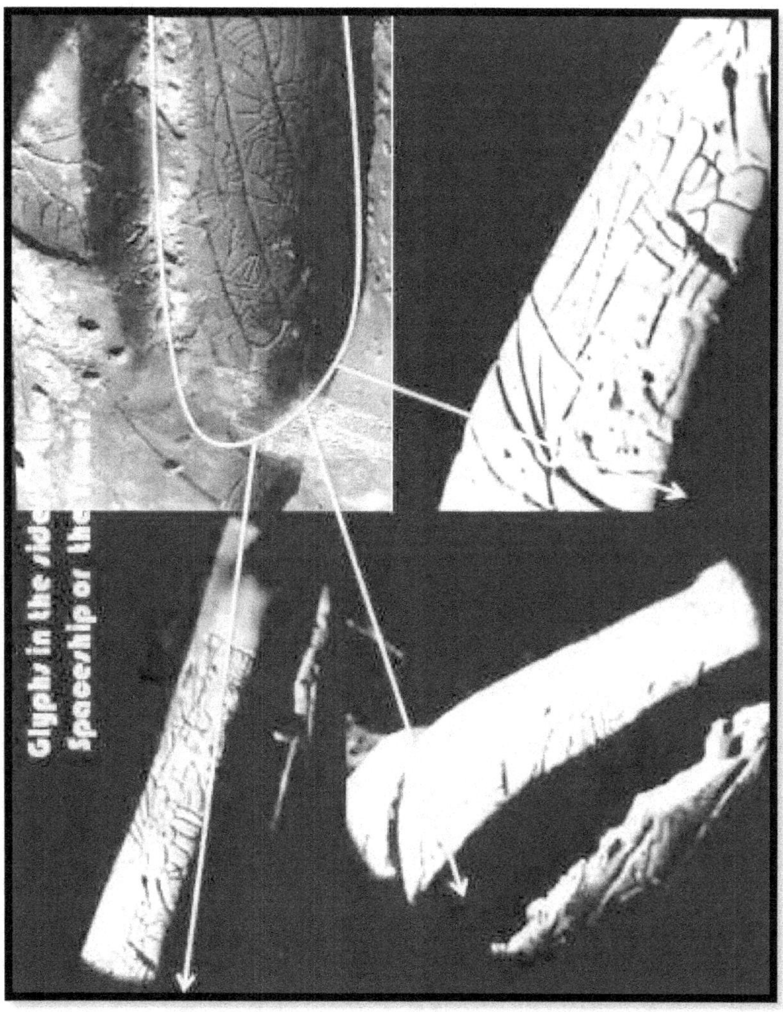

Belgium in 1930 and employed by USAF as test pilot on various aircrafts.
Based on his declarations, this mission considered the most successful one of the Apollo programs.

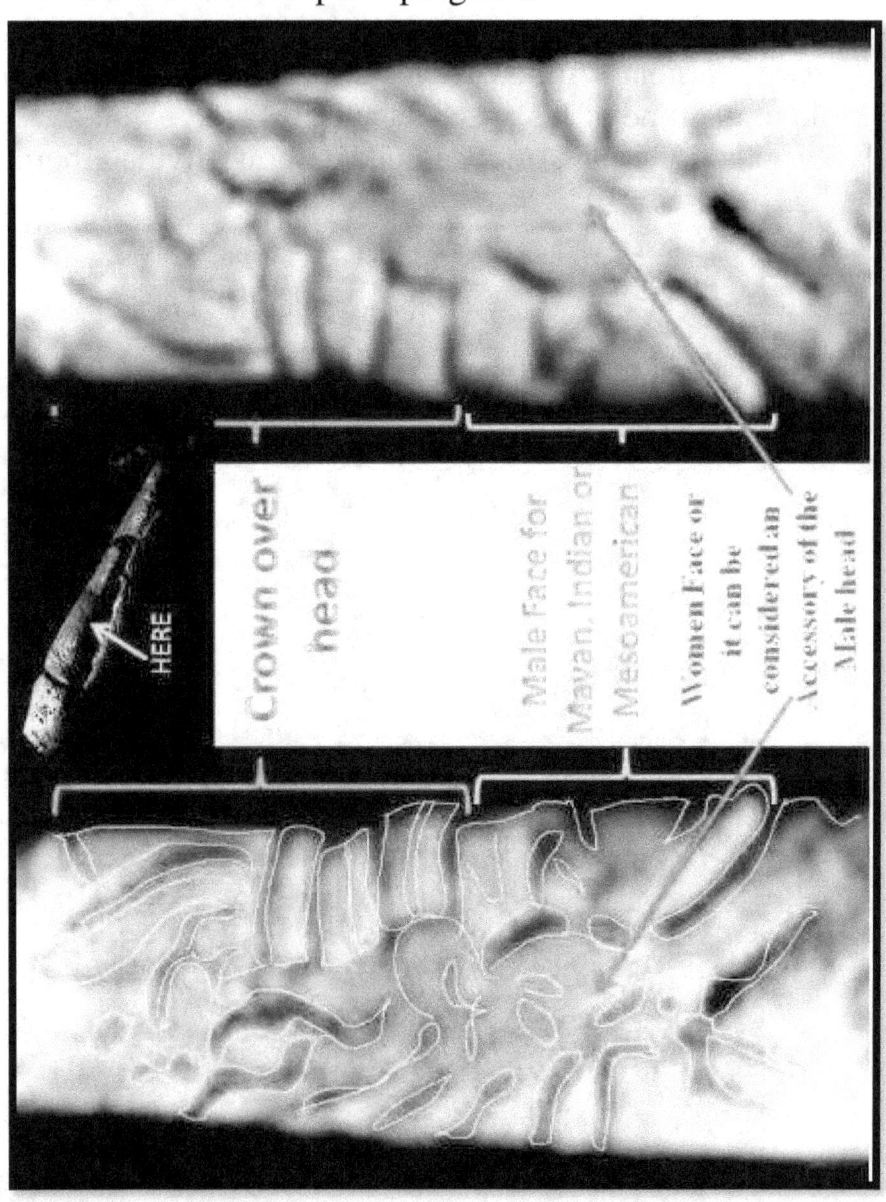

According to his statements , after the Apollo 17 program in December 1972 and the "Apollo18-Soyuz" mission taken place in July 1975, there were other two missions on the Moon: the Apollo 19 and the Apollo 20 (August 1976), which were both classified Space missions and launched from the Vandenberg Air Force Base in California.

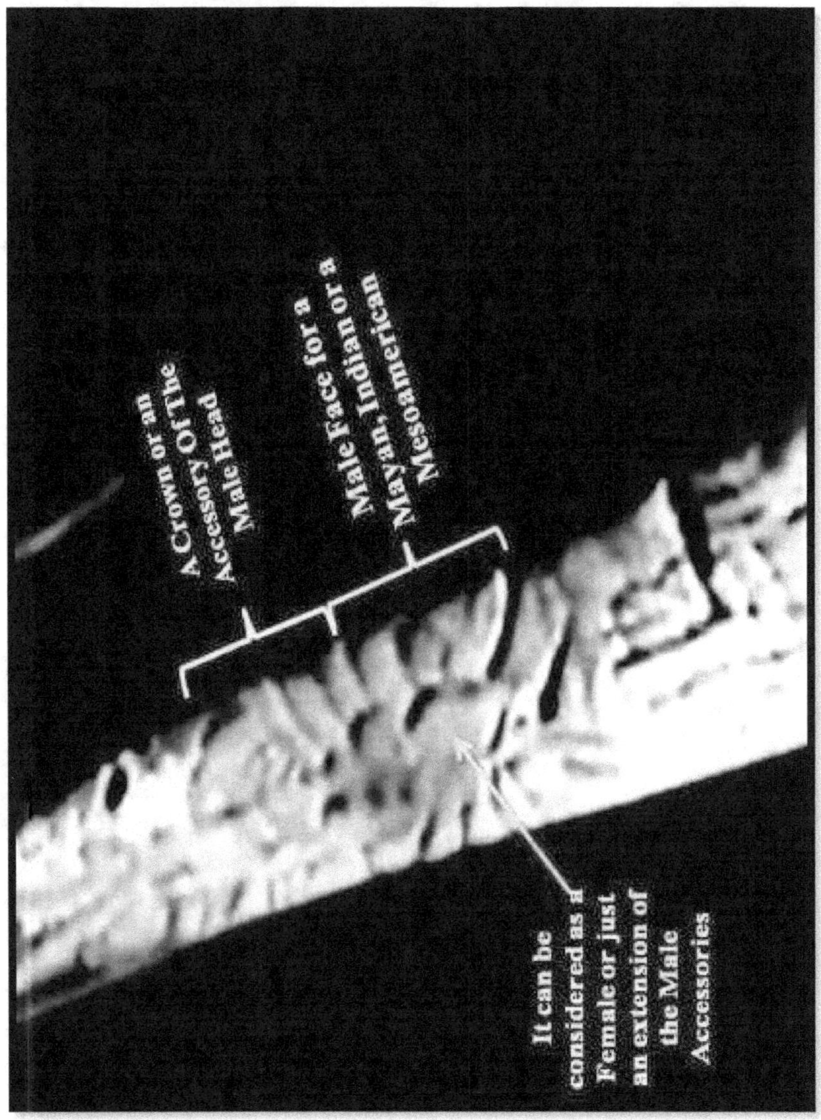

Officially many Apollo missions were canceled by NASA during the Project Apollo, included the Apollo 20 who was canceled in 1970.

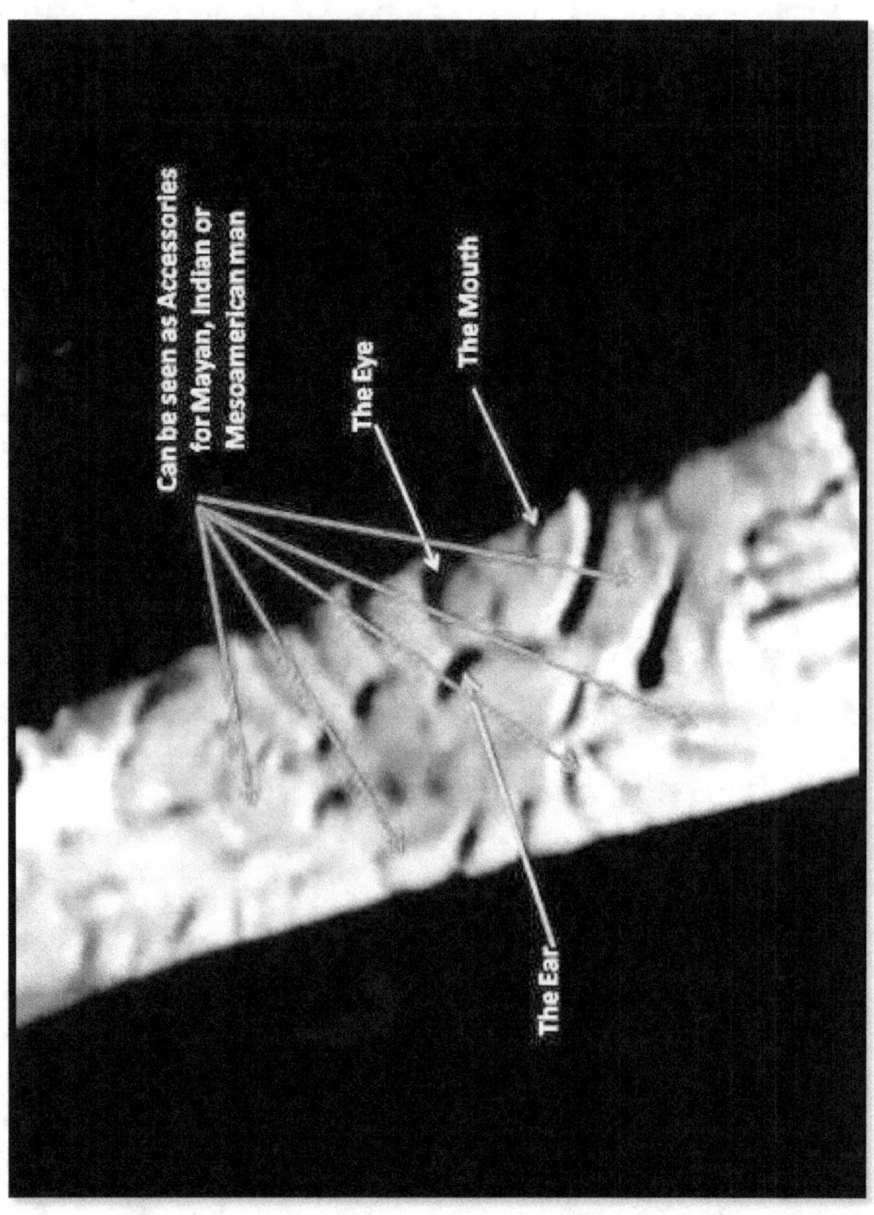

The goal of these two presumed secret joint space mission, result of an American-Soviet collaboration, was to reach the dark side of the Moon (The Delporte-Izsak region, close to the well-known Tsiolkovsky crater) and to explore a huge object found out during the Apollo 15 mission.

What the Apollo 20 crew found, it was a huge and ancient alien spaceship as he confirmed and as a matter of fact, some official NASA pictures archived by the LPI (The Lunar and Planetary Institute in Houston), which is "a research institute that provides support services to NASA and the planetary science community" show a strange and big object on the far side of the Moon. LPI is "managed by the Universities Space Research Association (USRA)".

The aerodynamism of the founding is near to be spaceship than a tower, but the draws founded in its surface with the material inside which not metallic, make the theory of tower is more evident.

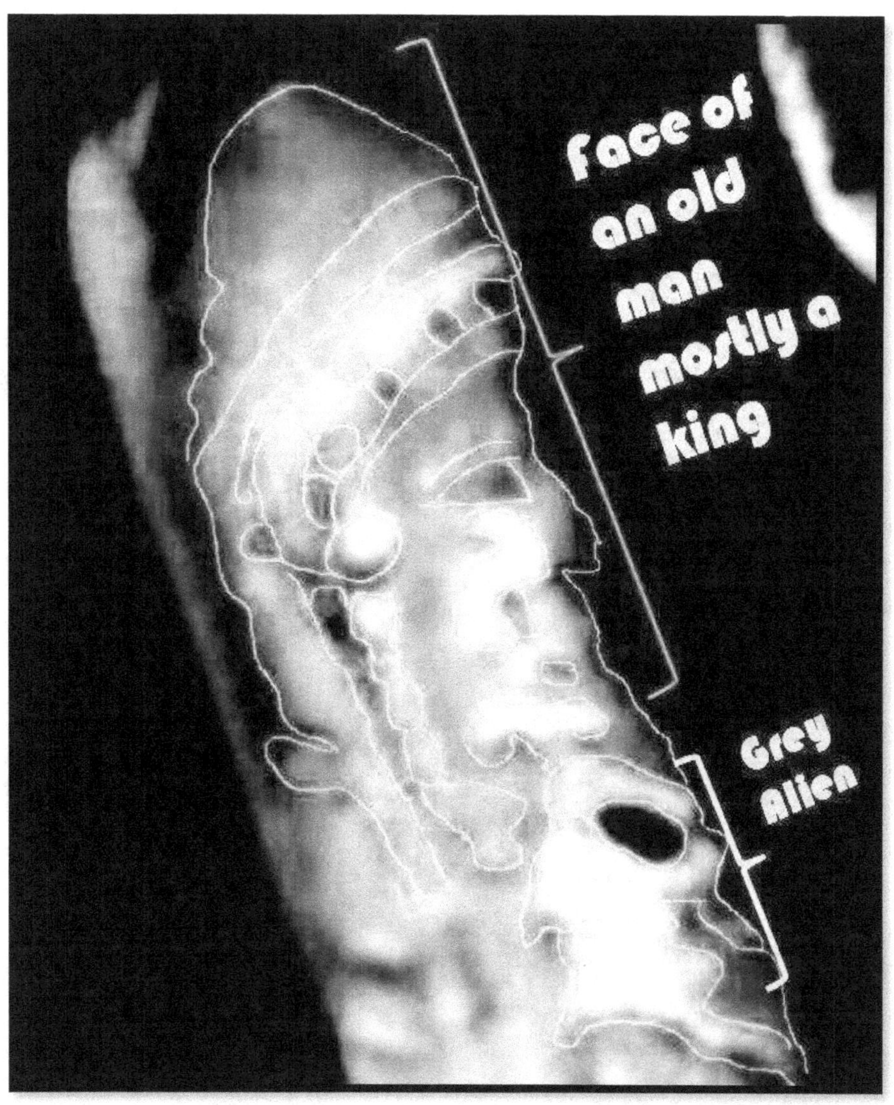

I discovered in the same side two other faces one for a
human, mostly an old man who can be a king or a priest
and the other is for a Grey Alien, in the top of the Mayan,
Indian or Mesoamerican male face:

This Spaceship or Tower was very clear when so many photos emerged. But after some time, it became covered with the moon soil. The logic question asked here: how it can be covered as none live on the moon?!!The answer is

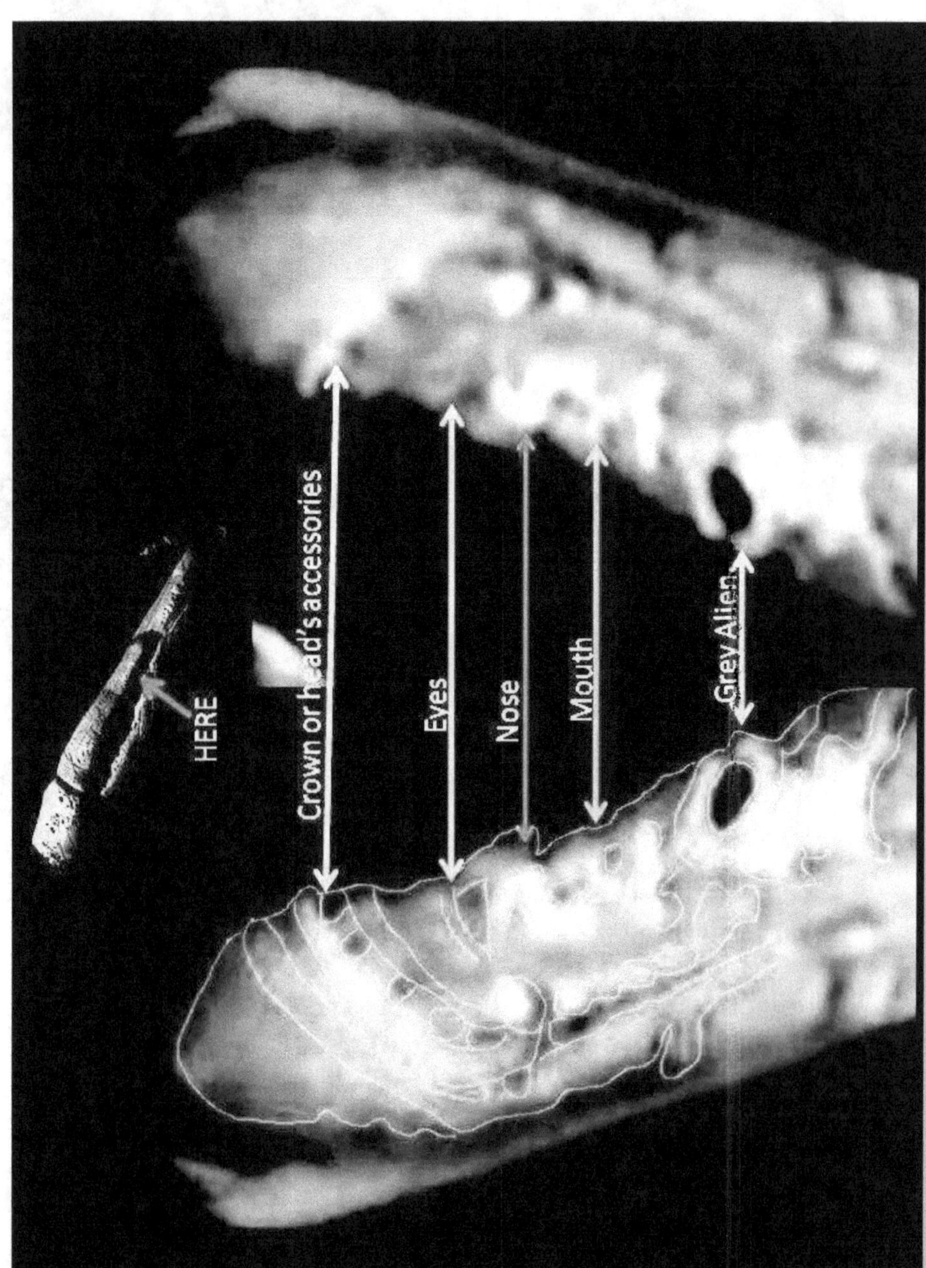

very simple, the moon is inhabited with several extraterrestrials beside Nazis, Americans, Russians and others. Daily, so many human made Flying Saucers go and came to and from the moon, meanwhile the mainstream media including the so called mainstream scientific media talk about the difficulty to go again to the moon and the moon has no atmosphere. The coverage of the Spaceship or Tower was made by workers there in the moon via the huge trucks, tractors and other engines.

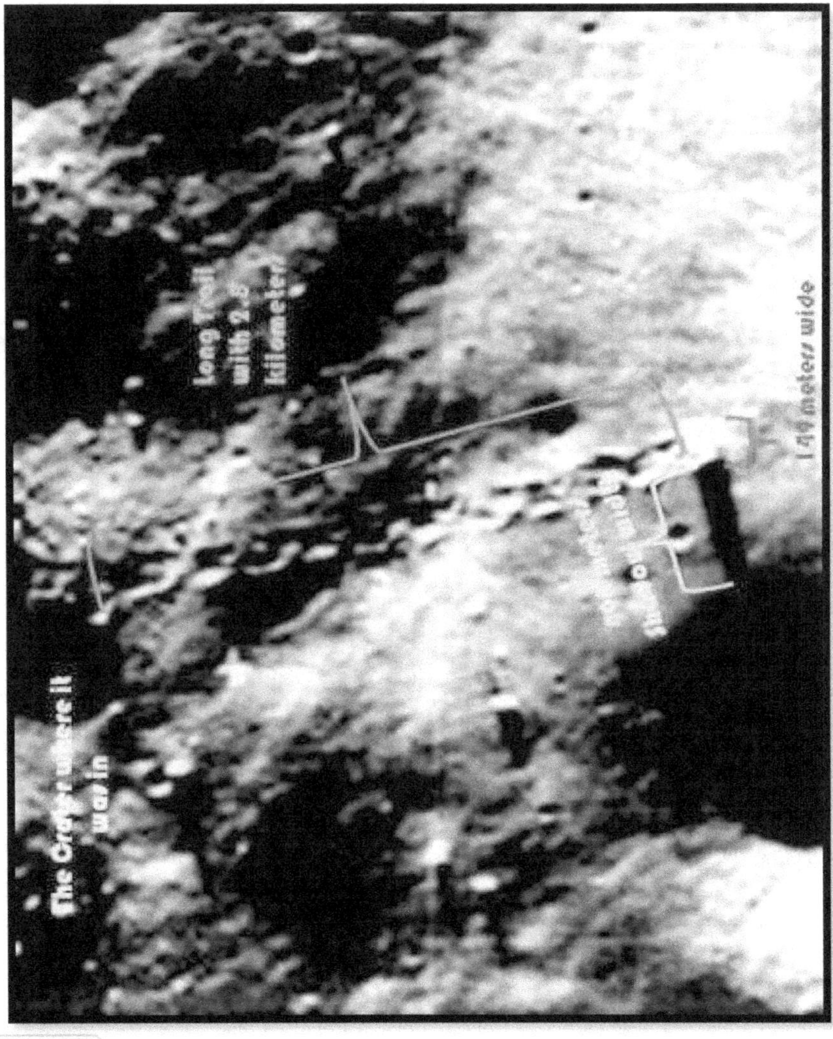

A huge mines truck was detected by Japanese Probe Selena and leaked to the public like other thousands of photos, from different Space Agencies, this truck left a crater where it was in, and continued to roll, leaving a long trail behind it with a distance of 2.8 kilometer.The width of the truck is 149 meters, leaving a height shadow of 490

meters. This explains the logic of the presence of towns and towers.

A huge concrete structure founded by the astronauts empowering this theory, comparing its size with the astronaut. Another factor do the same influence to this theory in moon is the existence of composition of the water: oxygen and hydrogen beside the nitrogen, important factors for the development of a life. Another secret discovery but in the same time anyone on earth can see it on moon photos is an energy matter: the Helium 3 which can replace oil and gas with a large difference in quality and quantity, it's used to construct huge structures and tours with the use of huge trucks this Helium three is located in the south pole of the moon, the light emerging is very strong.

This source of energy is used by the inhabitants of the moon, from some races of extraterrestrials, the Nazis, Americans, Russians and Japanese, each of them has his proper territories which defending it and fights any external intruders.

A huge Canon placed on the Top of artificial basis. This arm is very similar to the Nazis' ones which were used in World War two; this reinforces the theory of the existence of Nazis in the back side of the moon.

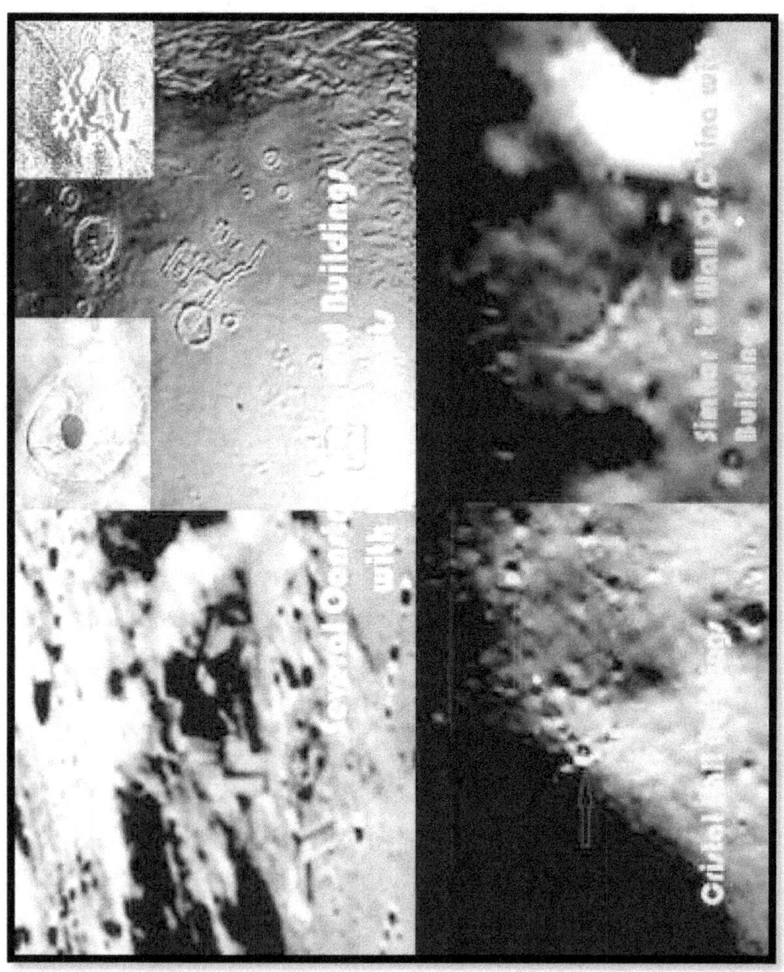

Other discoveries on the moon, is the presence of mini-spacecraft that's accompanied the astronauts from time to time in their walks, also a kinds of sensors connected to the ground of the moon that illuminates and send visual signals of light when astronaut approaches them, it is clear

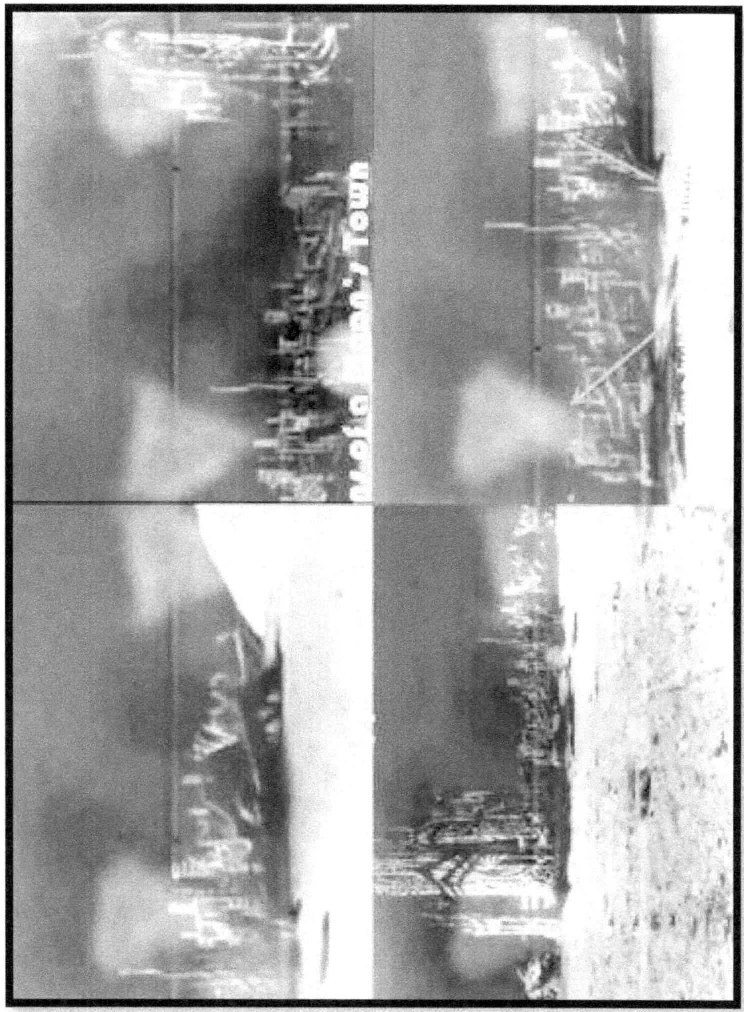

that they are security sensors alerting the presence of a stranger in these areas and borders.

Pictures taken from the center of spatial studies in Houston and Goddard Space Studies center with a huge tower made of transparent material surrounded by small cubes which are luminescent. We have here two potential scenarios that are can be imposed: the first: This crystal palace and its neighbors, they already existed and NASA has erased then,

the second, they did not already exist and have been built there and then it's also, proves that there is a continuous activity to build on moon. Another "Crystal Palace or Ball", with a form of inverted great radar, beside it there are others great structures with homes, A huge construction of walls like the great wall of china, also contains homes.

.

The supposed astronauts brought with them three bodies for three persons for a civilization which can be an Indian Americans, Mayans or Mesoamerican, and that's clear from their faces. The three bodies were well conserved. The bodies are covered with the wax matter this brings me to say that the three bodies are fake and made by this matter or the conservation itself was made by this matter.

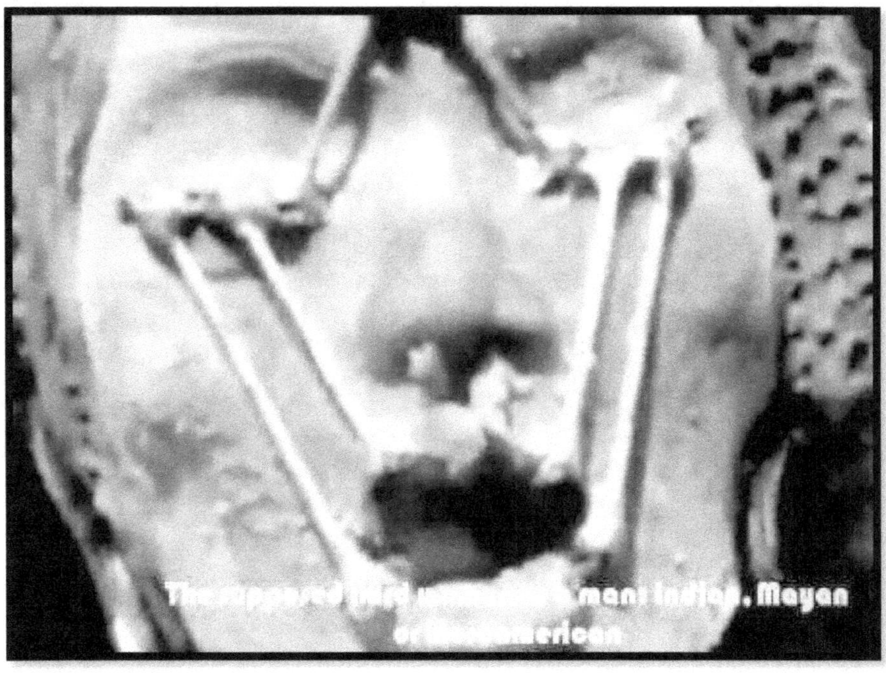

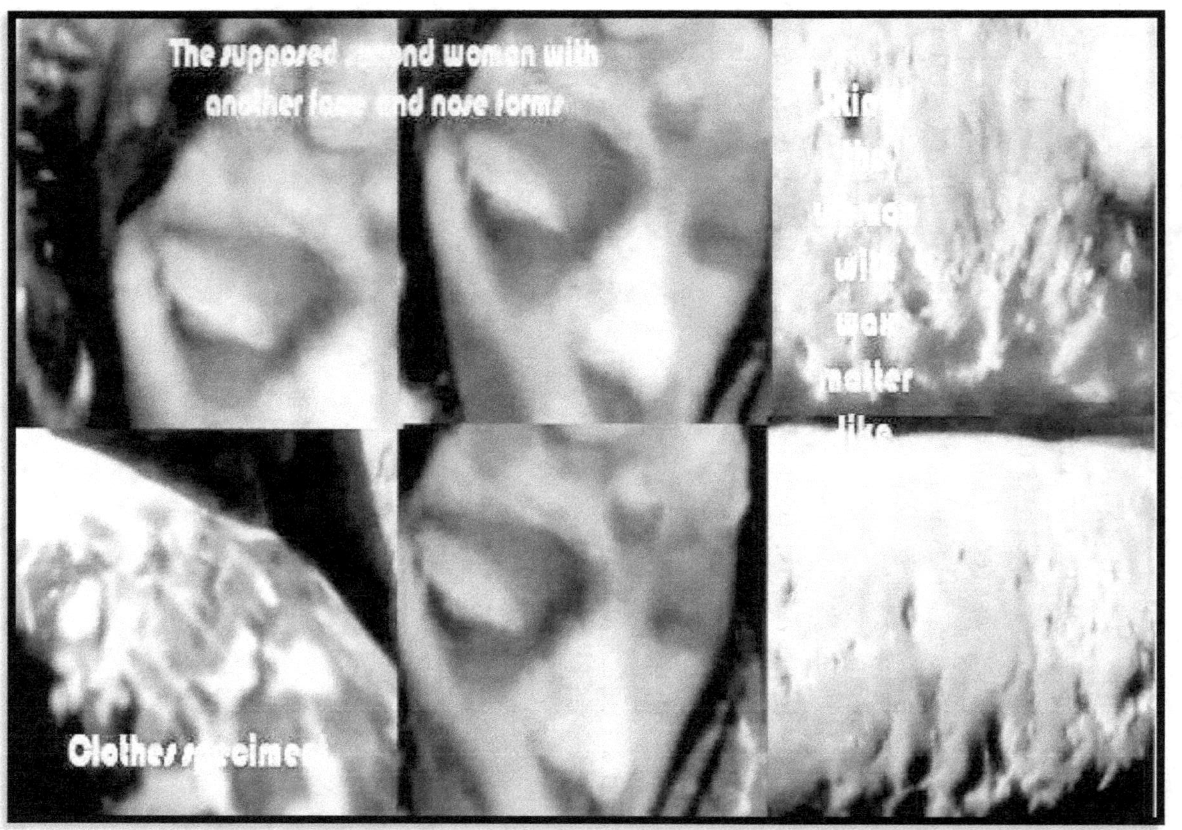
The supposed second woman with
another face and nose forms
if
will
matter
like
Clothes specimen

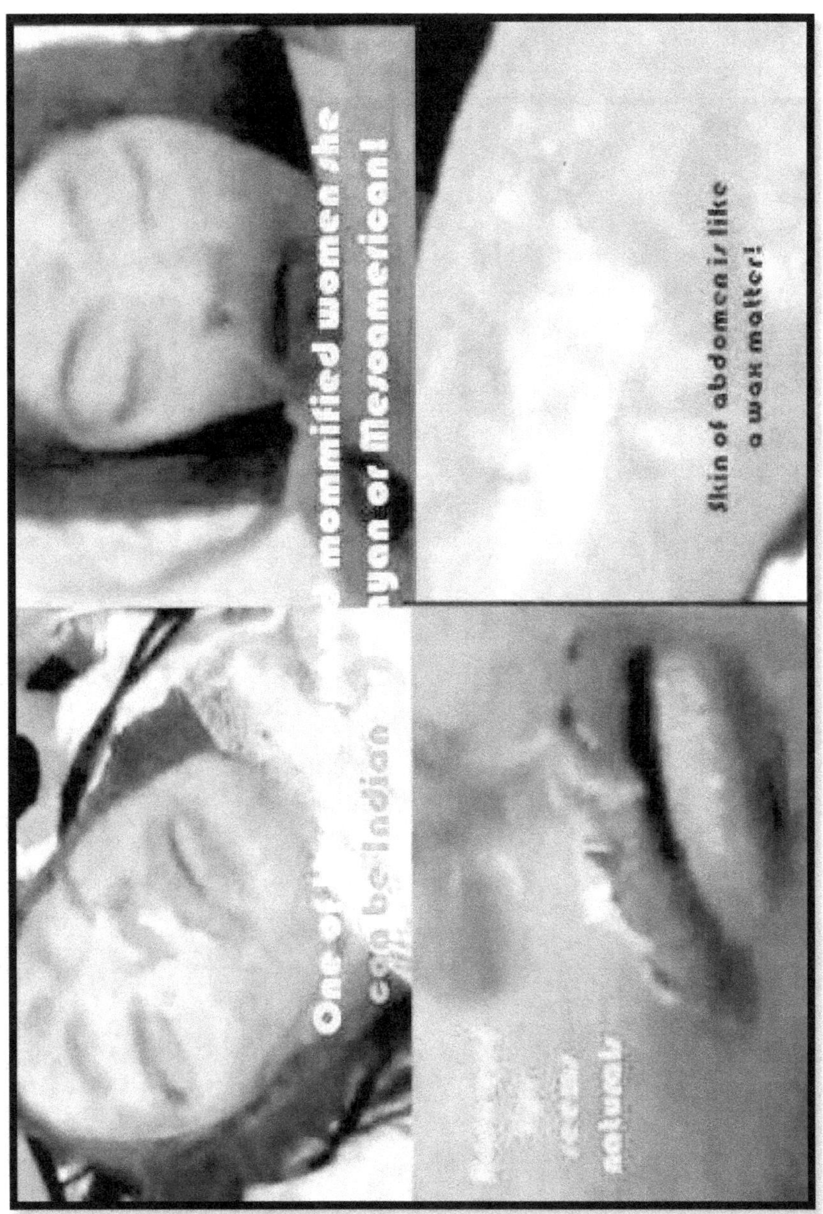

They bring with them written papers with strange language which appears like an inverted language:

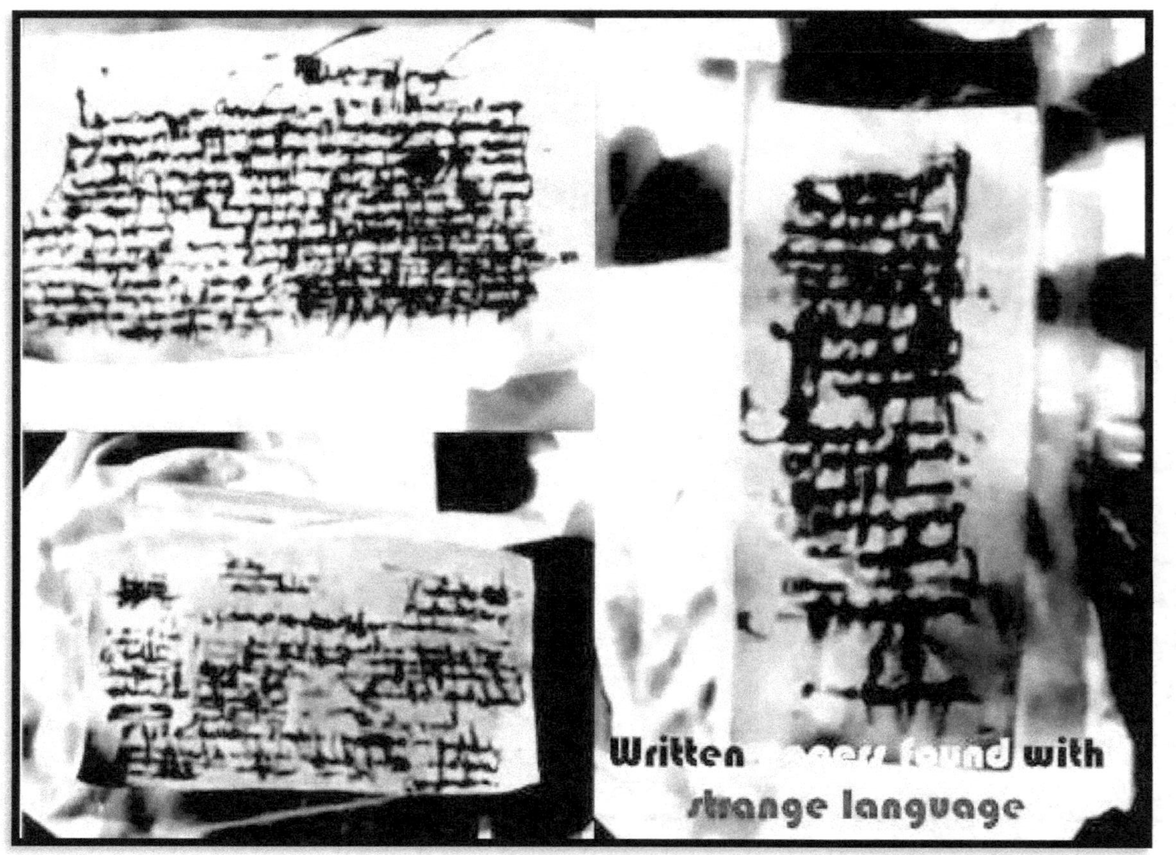
Written   ...ers found with
strange language

ABOUT THE AUTHOR

Mohamed Cherif focus was and still the extraterrestrials and their technologies. With time and continuous research he discovered that this subject is directly related with the world's Elites and their hunger to the control. Writes in international politics. Four years university studies Degree in Materials Physics Sciences